ROBOTS AND TELECHIRS:

Manipulators with Memory;
Remote Manipulators;
Machine Limbs for the Handicapped

M. W. THRING, M.A., Sc.D., F.Eng.

Professor of Mechanical Engineering
Queen Mary College, University of London

ELLIS HORWOOD LIMITED
Publishers · Chichester

Halsted Press: a division of
JOHN WILEY & SONS
New York · Brisbane · Chichester · Ontario

First published in 1983 by

ELLIS HORWOOD LIMITED
Market Cross House, Cooper Street, Chichester, West Sussex, PO19 1EB, England

The publisher's colophon is reproduced from James Gillison's drawing of the ancient Market Cross, Chichester.

Distributors:

Australia, New Zealand, South-east Asia:
Jacaranda-Wiley Ltd., Jacaranda Press,
JOHN WILEY & SONS INC.,
G.P.O. Box 859, Brisbane, Queensland 40001, Australia

Canada:
JOHN WILEY & SONS CANADA LIMITED
22 Worcester Road, Rexdale, Ontario, Canada.

Europe, Africa:
JOHN WILEY & SONS LIMITED
Baffins Lane, Chichester, West Sussex, England.

North and South America and the rest of the world:
Halsted Press: a division of
JOHN WILEY & SONS
605 Third Avenue, New York, N.Y. 10016, U.S.A.

© 1983 M. W. Thring/Ellis Horwood Ltd.

British Library Cataloguing in Publication Data
Thring, M. W.
Robots and telechirs. – (Ellis Horwood series in engineering science)
1. Automation
I. Title
629.8'92   TJ211
Library of Congress Card No. 83-10685
ISBN 0-85312-274-1 (Ellis Horwood Ltd. – Library Edn.)
ISBN 0-85312-651-8 (Ellis Horwood Ltd. – Student Edn.)
ISBN 0-470-27465-4 (Halsted Press)
Typeset in Press Roman by Ellis Horwood Ltd.
Printed in Great Britain by Butler & Tanner, Frome, Somerset.

ROBOTS AND TELECHIRS:
Manipulators with Memory; Remote Manipulators; Machine Limbs for the Handicapped

ELLIS HORWOOD SERIES IN ENGINEERING SCIENCE

STRENGTH OF MATERIALS
J. M. ALEXANDER, University College of Swansea.
TECHNOLOGY OF ENGINEERING MANUFACTURE
J. M. ALEXANDER, University College of Swansea, G. W. ROWE and R. C. BREWER, Imperial College of Science and Technology
VIBRATION ANALYSIS AND CONTROL SYSTEM DYNAMICS
C. BEARDS, Imperial College of Science and Technology
STRUCTURAL VIBRATION ANALYSIS
C. BEARDS, Imperial College of Science and Technology
COMPUTER AIDED DESIGN AND MANUFACTURE 2nd Edition
C. B. BESANT, Imperial College of Science and Technology
THE NATURE OF STRUCTURAL DESIGN AND SAFETY
D. I. BLOCKLEY, University of Bristol.
BASIC LUBRICATION THEORY 3rd Edition
A. CAMERON, Imperial College of Science and Technology
STRUCTURAL MODELLING AND OPTIMIZATION
D. G. CARMICHAEL, University of Western Australia
SOUND AND SOURCES OF SOUND
A. P. DOWLING and J. E. FFOWCS-WILLIAMS, University of Cambridge
MECHANICAL FOUNDATIONS OF ENGINEERING SCIENCE
H. G. EDMUNDS, University of Exeter
ADVANCED MECHANICS OF MATERIALS 2nd Edition
Sir HUGH FORD, F.R.S., Imperial College of Science and Technology, and
J. M. ALEXANDER, University College of Swansea.
ELASTICITY AND PLASTICITY IN ENGINEERING
Sir HUGH FORD, F.R.S. and R. T. FENNER, Imperial College of Science and Technology
LINEAR ANALYSIS OF FRAMEWORKS
T. R. GRAVES-SMITH, University of Southampton
INTRODUCTION TO LOADBEARING BRICKWORK DESIGN
A. W. HENDRY, B. A. SINHA and S. R. DAVIES, University of Edinburgh
THE MASONRY ARCH
J. HEYMAN, University of Cambridge
ANALYSIS AND DESIGN OF CONNECTIONS: Reinforced Concrete and Steel
M. HOLMES and L. H. MARTIN, University of Aston in Birmingham
TECHNIQUES OF FINITE ELEMENTS
BRUCE M. IRONS, University of Calgary, and S. AHMAD, Bangladesh Unversity, Dacca
FINITE ELEMENT PRIMER
BRUCE IRONS and N. SHRIVE, University of Calgary
PROBABILITY FOR ENGINEERING DECISIONS: A Bayesian Approach
I. J. JORDAAN, University of Calgary
CONTROL OF FLUID POWER, 2nd Edition
D. McCLOY, Ulster Polytechnic, N. Ireland and H. R. MARTIN, University of Waterloo, Ontario, Canada
TUNNELS: Planning, Design, Construction
T. M. MEGAW and JOHN BARTLETT, Mott, Hay and Anderson, Consulting Engineers
UNSTEADY FLUID FLOW
R. PARKER, University College, Swansea
DYNAMICS OF MECHANICAL SYSTEMS 2nd Edition
J. M. PRENTIS, University of Cambridge.
ENERGY METHODS IN VIBRATION ANALYSIS
T. H. RICHARDS, University of Aston, Birmingham.
ENERGY METHODS IN STRESS ANALYSIS: With Intro. to Finite Element Techniques
T. H. RICHARDS, University of Aston in Birmingham
COMPUTATIONAL METHODS IN STRUCTURAL AND CONTINUUM MECHANICS
C. T. F. ROSS, Portsmouth Polytechnic
ENGINEERING DESIGN FOR PERFORMANCE
K. SHERWIN, Liverpool University.
STRUCTURAL DESIGN OF CABLE-SUSPENDED ROOFS
K. SZABO, Budapest Technical University and L. KOLLAR, City Planning Office, Budapest
ROBOTS AND TELECHIRS
M. W. THRING, Queen Mary College, University of London
NOISE AND VIBRATION
R. G. WHITE, J. G. WALKER, University of Southampton
STRESS ANALYSIS OF POLYMERS 2nd Edition
J. G. WILLIAMS, Imperial College of Science and Technology

Table of Contents

Dedication

To my wife, who has taught me over more than forty years how infinitely superior a human being is to the most perfectly conceivable robot.

Preface

When I was a small boy in the twenties, my fancy was caught by magazine articles about wonderful manlike robots and, as electronics had not yet been developed beyond the 'wireless' crystal and triode valve, I designed robots to work by mechanical devices such as the sound responsive disc of the phonograph.

Thirty years later, in the fifties, when I was Professor of Fuel Technology at Sheffield University, I came back to the study of robots. In the meantime, I had learnt thermodynamics as part of my Physics degree, so I knew about the need to provide an adequate source of power, and electronics had developed by then as far as the first computers and transistors. My renewed interest started from my observation of how hard my wife worked to maintain a good home, with three young children and an absent-minded husband! Much of this work was repetitive drudgery so I began studies of the possibility of a domestic robot which would take over this drudgery. I continued these studies all through the sixties, during which period I became a Professor of Mechanical Engineering at Queen Mary College of London University. I extended these studies to the design of prototype robots for fire-fighting, storekeeping, night-watching and assembly work.

Three vitally important problems have gradually, during the seventies, caused me to decide that an engineer with an active conscience should direct his work to rather different objectives.

(1) As part of my work on the domestic robot I developed a stair-climbing carriage and in doing so realised that this was needed for the much more urgent and vital problem of use by handicapped people. This led me to the study of Sceptrology (mechanical crutches) and to the establishment at QMC of an engineering group making devices in this field and novel equipment for helping doctors, surgeons, physiotherapists and nurses. I have come to feel, as have many other engineers, that this field of Engineering is one that is totally satisfying to ones conscience; although one for which money is tragically difficult to find.

(2) Over the last thirty years I have seen the steady rise of unemployment in the developed countries as they reach the inevitable end of the period of continual

economic growth. As an engineer it is clear to me that such growth, based as it is on a continual growth in the use of limited mineral resources, must be replaced by *a steady state* or *equilibrium* economics. I have gone into this matter in detail in my book *The Engineer's Conscience,* but here I am concerned only with the consequence as far as our use of robots is concerned. My conclusion is definitely that from the long-term point of view (the life of my grandchildren in the next century) the use of robots purely to displace human workers, because robots are cheaper to the employing firm, is potentially disastrous. If, of course, the displaced humans were provided with another more interesting job, which was appropriate to an equilibrium world economics, then the situation would be completely reversed. In that case, the robots become as beneficial to humanity as are the robots that free humans from dangerous, tiring or uncomfortable jobs. This is why I have shifted my personal studies from robots to telechirs (especially for mining) since mining telechirs offer an unequivocal benefit to humanity: the ultimate possibility of providing us with all solid minerals (even those inaccessible to human miners) as easily and safely as we win oil and gas at present.

(3) In 1963 I gave a paper on the idea of a domestic robot to the Royal Society of Arts, and Sir Harold Hartley, in his chairman's remarks said, 'I felt a little doubtful about his saying that the object is not raising the standard of life but creating more happiness, because, after all, what he was thinking of was this country and the United States, and there are so many other countries where the first duty of applied science is to try to raise the standard of living'. This remark brought home to me the vital world problem of the ever-increasing millions of people living far below the poverty line, and three years later in my Cantor Lectures to the Royal Society of Arts I said, 'The more developed countries have a moral responsibility to find a technological way of aiding the less developed ones to achieve a satisfactory standard of living for all their peoples. The peace of the world depends on this'.

In 1979 I was a member of a ten-man Unesco Commission to study ways in which technology could be applied in Bangladesh. I saw such results of poverty in towns, villages and hospitals that I came away with an intense feeling that the studies of high technology in the developed countries are concerned with solving problems either of relative triviality or even of real menace to my grandchildren.

It is absolutely clear to me that unless the rich nations soon begin to devote a substantial fraction of their engineering resources to assist the poor nations to become properly self-supporting, there is no hope whatever that my grandchildren will live in a peaceful world in the twenty-first century.

ROBOTS, TELECHIRS AND SCEPTRES
The distinction between the three kinds of machine with man-like limbs lies in the purposes for which they are built and in the mechanism of control.

A Robot is a machine which can be programmed by a human master to carry out any one of a considerable variety of complex manipulations which has to be repeated many times; its gripping device ('hand') is moved under the control of the master through a series of movements of grasping, positioning and orienting. It then stores the information about these movements in a computer or other mechanical information storage system and repeats them over and over again. It may or may not have the ability to modify these movements as instructed, according to tactile or optical sensory information received during each task. It may have the chassis fixed in one place and only move the the arms and hand, or it may be self-propelled and self-steered in a working region. Robotics is the technology of robots.

A telechir ('distant hand') is a machine which carries out the manipulative instructions of a human master who is connected to it all the time it works by mechanical linkage, cable, optical fibre or short-wave radio. The master must receive sufficient visual, and possibly also tactile, information to control the hands of the telechir to do any task that he would do directly if he were on the spot. Thus each job it does can be quite different from the one before, and it makes use of the adaptability and initiative of a fully trained human who can deal with quite unexpected situations. Its function is to enable the human to do his work in positions which he cannot reach with his body for reasons of safety, comfort, accessibility or insufficient space. Telechirics is the technology of telechirs.

A sceptre (Greek word meaning either a king's staff of power or a crutch) is any mechanical device to assist a person with handicapped arms, hands or legs. It must be controlled all the time it is in use by the handicapped person, or his nurse, to achieve his purpose. Sceptrology is the technology of mechanical aids for people with handicaps of the limbs.

ACKNOWLEDGEMENTS

My old friend Professor W. B. Heginbotham has helped me in many ways with the preparation of the manuscript especially by letting me have a complete set of the numerous and cogent publications describing his work in the area of Robotics.

I have had many fruitful discussions over the years with my friends on the CISM-IFToMM Symposia Organising Committee, especially Professors Bianchi, Kato, Konstantinov, Morecki and Roth and Doctors Vukobratovic and Vertut.

Mrs Audrey Hinton has helped all the way through the preparation and typing of the manuscript.

Mr K. Siva has done a great deal of work on the figures and the theoretical treatment.

CHAPTER 1

What does man want from slave machines?

1.1 MAN'S WANTS AND NEEDS

Ancient civilisations were based very largely on the use of human slaves to do
those tasks which the rich and powerful owners of the slaves did not wish
to do. The civilisation of the countries which have the benefits of the industrial
revolution is based essentially on the idea that we use machines, powered by
fossil fuels, both to do those tasks which we do not wish to do and also to enable
us to do things (like rapid travel, or hearing and seeing remote performances)
which were not available to our ancestors and to earlier civilisations. On the
other hand, there has always run through civilisations the same religious idea
that hard work and even menial work was necessary for the proper inner self-
development of an individual.

The development of the computer and the microprocessor have led to the
possibility that we could develop mechanical slaves to free a substantial fraction
of the world's population from the manual work connected with the production
of food and manufactured goods. It is true that after the first generation of
'senseless' robots has been on the market for 20 years there are still only about
15,000 of them in the whole world, more than half of this number being in Japan,
whereas the number of manual workers engaged in production of food and
manufactured goods in the world is more than 1000 million, so that there is
only one robot to every 100,000 people. It is also true that robots require a
great deal more energy[†] (admittedly supplied in the form of fossil fuel rather
than food) than a human doing the same job, and they require both fossil fuel
and other limited resources such as metal ores to make them. It is probable
that these limitations will mean that there will never be more than a few hundred
thousand robots in the world.

[†]The energy equivalent of the food of a normal human is approximately 3000 kcal/day or
0.16 tons of coal equivalent/year or 140 W. The energy output of a man engaged in hard
manual labour is less then $\frac{1}{4}$ HP = 180 W. A robot will consume at least 1 kW and more
often 5 kW when working.

So for two reasons we have to consider very carefully what are the most appropriate fields of application of robots and other humanoid machines.

(i) Because of the limited resources of fossil fuels and metal ores available to the very large number of people on the earth and the still greater number that will arise in the future.

(ii) Because there may be something in the idea of the value of physical work to human beings.

This latter question also relates to the common and well-justified fear that robots will throw humans out of work. Indeed the economic reason that most factories employ robots is that they are cheaper than humans for the same job, especially if they work two shifts or even three. From the human point of view the idea of using a robot to throw a person out of work is clearly highly unsatisfactory since normal human beings would much rather do a job than be paid money for doing nothing, provided they feel the job is worth doing. Moreover, it costs the country so much to pay them for doing nothing that on a national basis the economic benefit to the firm is totally destroyed.

1.2 THE TWO POSSIBLE FUTURES FOR MANKIND

In my book *The Engineer's Conscience* I have developed the theme that, if we continue to decide what machines we shall develop and use purely on the short-term and national bases of maximising the standard of living and the weaponry of each separate nation, we shall run into disaster in one or another way. The two disasters which are particularly relevant to our discussion of the possible uses of robots and telechirs are as follows.

(i) First there is the extreme danger of world war resulting from the tensions caused by the increasing poverty of millions of the world's inhabitants who exist in the same world as other millions who are short-sightedly exhausting the earth's limited resources of fossil fuels (especially oil) and the richer and more accessible deposits of mineral ores such as ferrous and non-ferrous metals and P & K (needed for fertilisers). The rich countries continually escalate their weapons although the great powers have long exceeded the number of weapons needed for the total destruction of each other's civilisation; they sell arms to anyone who can pay for them and the number of countries with nuclear bombs steadily rises. The danger is not that the poor countries will attack the rich ones but that the sheer frustration of poverty combined with the realisation of the richness of other parts of the world will cause a fertile soil for the revolutions, civil wars, and local wars which will gradually involve the military great powers and bring them eventually into military opposition to each other. Such situations are incipent in many parts of the world at the present moment.

All attempts to halt the build-up of nuclear arms by military great powers or the spread of nuclear bombs to more nations are proving un-successful, so that the elimination of this world tension in the next 30 years is clearly in the vital interests of the rich nations if they do not wish to be destroyed by the use of nuclear weapons. This means that for their own survival they must gradually devote their most powerful technological developments, not to further weaponry nor to greater and greater sophisti-cation, but to the basic problem of giving all the people of the world a decent standard of living and the other positive benefits of the industrial revolution within the earth's limited resources.

(ii) The second disaster will come from the increase of unemployment in the rich countries to a level at which the peaceful fabric of their society breaks down because so many people find their services and skills rejected by society. As long as the extra productivity of the worker resulting from machines could be absorbed by increasing the general standard of living and the use of fossil fuels and raw materials these new machines did not cause unemployment. However a 4% per annum increase in workers' pro-ductivity means a 50-fold increase in a century and hence must be absorbed by a 50-fold increase in the consumption of goods and at least a 10-fold increase in energy and metal ores if escalating unemployment is to be avoided. Since we are already pushing against the limits of these minerals and the poor countries have not begun to receive the benefits of the industrial revolution, such an increase of consumption is no longer possible. It is not possible to reduce the working hours to $\frac{1}{10}$ of their former value since most people need a regular work framework for their lives. One is therefore forced to the conclusion that further increases in workshop productivity given by the engineers' development of robots, automation and mechanisation must cause steadily rising unemployment. A careful analysis of the needs of normal people certainly indicates firstly, that they need to feel that society finds their services useful and does not just pay them a dole for doing nothing; secondly, that they need the framework of a regular working week — very few can provide the self-drive of an author working at home all her or his life; and, thirdly, that while everyone would like an interesting job that uses their skills of head, heart and hands to the full, if this is not available many people would rather have a job of low interest for the above two reasons than to be unemployed.

Taking account that we must avoid these two disasters if our civilisation is to survive through the twenty-first century we can list the potential advantages and disadvantages of further development of robots, artificial intelligence, automation, sceptrology and telechirs as follows.

Potential Good Results
G.1 Helping handicapped people to live a more normal life (sceptrology, such as powered artificial arms and legs, powered carriages).

G.2   Elimination of danger and discomfort at work (robots and telechirs).
G.3   Increase of available resources and reduction of wastage; recycling (telechirs, automation).
G.4   Avoidance of drudgery and boring repetitive work (robots, artifical intelligence, automation).
G.5   Increase of human knowledge of the Universe and of man's significance in it (artificial intelligence, robots and telechirs in space).

Potential Bad Results

Unfortunately, however, there are also five potentially bad results of an unthinking pursuit of sophisticated technology in these directions resulting from three short-sighted motives as follows.

M.1   The pursuit of immediate monetary profit.
M.2   Solving problems for their excitement and interest regardless of their consequences.
M.3   War preparations.

The potentially bad results of sophisticated machines are as follows.

B.1   *Increased unemployment* in the developed countries. If in the long term we are to have a job for everyone, then the short-term objective of using robots because they are cheaper than humans is *in itself* a move away from a stable society. Moreover, it is an economic fallacy (which we can call the *robot fallacy*) in most cases because the total saving to the nation when we include the loss of workers' income tax and the payment of unemployment pay becomes negative. When therefore it is decided to use robots for some of the good reasons given above this contributes to the general problem of providing an interesting and worthwhile job for everyone in a mechanised society which is discussed in detail in Chapter 7 of my book *The Engineer's Conscience.*

B.2   *The excessive use of limited resources* (e.g. fossil fuels, oil, coal, natural gas, metal ores, P & K) animal or fish species destruction, despoliation of land (by concrete and monoculture) and fresh-water depletion.

B.3   Putting humans into jobs where they are more isolated or do less interesting work (e.g. computer programmers and assembly belt workers).

B.4   Pollution and danger to public health.

B.5   Accidents to humans caused by sophisticated machines.

Many people when considering the posibilities of robots have envisaged a future for mankind in which all physical tasks are done by robots and the whole of these activities is operated and maintained by a relatively small number of technocrats working 10 or 20 hours a week. On this world picture the rest of humanity is either unemployed or engaged in a human service activity.

This world view is totally incompatible with both the limitations of raw materials, agricultural land, water and fresh air and equally is totally inconsistent with the essential characteristics of human nature. Every human being has a lazy side which would lead him to prefer to make no efforts at all and the traditional weapon used by society to overcome this laziness is the simple economic one of no food without work, and it is remarkable how necessary the economic motive coupled to pride in one's possessions is to effective and hard work in the two cases of growing crops and looking after one's home. The best examples of the really efficient use of land are those where the farmer or smallholder combines improvement of his own economic position with the pride of achievement, similarly houses are most economically maintained and improved when the owner-occupier is personally responsible for them. The ordinary human being also has more subtle motives such as the desire to earn his or her living by something which they feel emotionally is of real value to other people whether on the small scale, as in the bringing up of children, or on the large scale, as helping to feed a starving nation. Another more subtle motive is the desire to express oneself through original creative work and yet another is the feeling that it is good for one's own self-development to struggle against one's laziness by wise effort.

To satisfy all these basic human motives requires an entirely different kind of society, a form of machine-served Utopia in which the machines are used only sufficiently to enable everyone to have a decent standard of living within the earth's resources and to earn this without excessive drudgery or repetition and without danger. In this book robots and telechirs and their development, both theoretical and practical, will be described with this kind of possible future world in mind, rather than the technocratic society or the use of robots and telechirs for purposes of destruction. This kind of society may be called a 'creative society' because the profit motive would be used only to the point of enabling everybody to earn enough of everything and their extra abilities and energies could be directed towards fulfilling their fundamental creative instinct.

Since we are concerned with man-like machines it is necessary to mention at this point one other important fact of observation about the human system. This is the fact that a human being has three separate brains[†] which work at different speeds and are educated by different means and have entirely different functions. These brains may work together, they may interfere with each other or they may be working entirely unconnectedly as when a human in a factory carries out some routine task with the hands and eyes, with the 'body brain' or 'physical brain', while there is some imaginary conversation going on in the 'head' or 'intellectual brain' and there is a seething in the background of depression, anger or boredom in the 'emotional brain'. The functions of the three brains

---

[†]This concept is clearly stated in the account of Gurdjieff's teaching given on p. 53 of *In Search of the Miraculous* by P. D. Ouspensky, 'he spoke of three centres, the intellectual, the emotional and the moving', (Harcourt Brace, 1949).

are listed in Fig. 1.1. All truly creative work requires the coming together of the three brains as discussed in my book, written in conjunction with Professor E. Laithwaite, *How to Invent*. The 'emotional brain' has to provide the tremendous power to leap into the dark away from accepted thinking while the 'intellectual brain' gives a logical analysis of the problem with as little rigidity as possible;

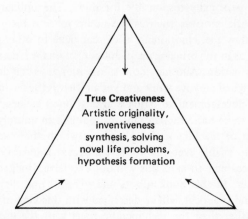

Emotional Brain

Symbol: the heart, solar plexus or pit of the stomach.

Judgements, conscience, motives, religious feelings, emotions, such as love, happiness, and sadness, human sympathy in relationships, values

True Creativeness
Artistic originality,
inventiveness
synthesis, solving
novel life problems,
hypothesis formation

Intellectual Brain
Symbol: head, cerebral hemisphere.

Concepts, logic, philosophy, mathematics, chess-playing, knowledge of scientific laws and principles, analysis, theory, numeracy, abstraction.

Physical Brain
Symbol: whole body, eyes, hands, legs.

Knowledge of reality through observations and manipulations, muscular sensing, craft skills, artistic techniques (painter, sculptor, musician, singer).

Fig. 1.1 — The three brains of a human.

and the 'physical brain' provides the essential direct knowledge of reality. After a new idea has been thought of, the critical faculty in the 'intellectual brain' is switched on to consider whether it is contrary to the known laws of physics, chemistry or medicine. Many inventions, such as the ability of an ordinary person to learn to balance a bicycle, would clearly have seemed to be contrary to experience until they were tested. The conversion of an idea into reality always involves the 'physical brain' with its experiential understanding of static space and force relationships, kinematics and kinetics.

1.3 THE HISTORICAL DEVELOPMENT OF MECHANICAL SLAVES

Before the development of sources of power, machines were mainly hand tools and simple devices such as the wheel, the wedge and the lever. Crude cog wheels and screws were used for water-lifting devices, powered by men or animals. Simple windmills and sailing ships were the first additional sources of power followed later by the water wheel. Windmills and water wheels rarely gave more than 10 h.p. During this period the robots which have been described in literature were either mechanical toys operated by clock springs, legends, such as the brass giant Talus, or fakes, such as the von Kempelen chess player which was operated by a dwarf inside. Many very beautiful and elaborate clockwork toys were produced, sometimes using the principles of the Swiss musical box with a drum rotated by clockwork with pins arranged along it to work different mechanisms as they came round. Such machines were, however, entirely limited to a few simple operations with a mechanical programme to carry them out.

The next stage can be described as the *first stage of real mechanisation* which was the introduction of sources of power capable of operating much larger and more effective machines than the hand tools so far available to man and with them the machine tools such as the accurate lathe with slide rests to replace the traditional treadle-operated lathe with hand-held tools. The power available to a single worker has steadily increased so that now a driver may be operating a 100 h.p. tractor or bulldozer compared with the 1/8th h.p. or so of a human navvy. The limit of the old water wheels and windmills was of the order of 10 h.p. and a large beam pumping steam engine built in Holland early in the nineteenth century by a Cornish firm had six pumps which replaced 150 windmills. In this stage of mechanisation a human operator is in control of any changes of operation the whole time.

The first example of the second stage of mechanisation was the device at the back of a large windmill which rotated the main blades to face the wind whenever there was a component of wind in the plane of these blades. This type of control is *simple automatic and sequence control.* A human can set it to do one kind of operation and leave it to carry out this operation. For example, one can have a device which will maintain the temperature of a furnace constant (by varying the fuel input) when a load is taken into or out of the furnace, or it may vary the temperature of the furnace over a pre-set time cycle. The system may contain a clock or a device for carrying out a sequence of operations, but in general there are only a very few variables being controlled. Examples of this stage of mechanisation are the programmed washing machine or dish water and the boiler with a thermostat. The first sequence control on a machine tool was the *automatic lathe,* developed early in this century, in which a large set of drums driven at a slow speed underneath would have variable cam pieces bolted to them to carry out in sequence a series of operations on a *turret lathe* by one complete rotation of the drums. This is now developed to the more sophisticated *numerical control* machine which, however, operates without feedback so that if the tool is worn it will no longer produce the object to the correct dimensions.

In the *third stage of mechanisation* come the robots and telechirs which are the subject of this book. They correspond much more closely to a human worker capable of using a variety of hand tools in different operations or of feeding work pieces into and out of machine tools or heat-treatment furnaces.

1.4 CLASSIFICATION AND DEFINITIONS

I shall use the classification in Fig. 1.2. The *first group* is mechanical limbs or mechano-chiropods directly attached to the human body. These can be divided into two sub-groups. The first is limbs to help people who have lost the use of arms, hands or legs to carry out more normal functions. For these I use the word *'sceptrology'* because the Greek word for sceptre refers not only to the king's staff of power but also to a crutch. These are discussed in Chapters 5 and 6. They may be placed in the same relation to the handicapped person's body as would a real arm or leg. Alternatively arms and hands, quite separately supported and controlled by movement of the patients's eyes or blowing, have been used to enable quadriplegics to do simple handling movements for themselves while legs may be replaced by wheels or other devices to give similar mobility. The exoskeleton is a device where a man is surrounded by a strong steel framework and he controls the powered movements of its arms, legs and hands by control systems operated by the movements of his own limbs. This enables him to pick up a weight considerably more than 10 times that which he could handle unaided.

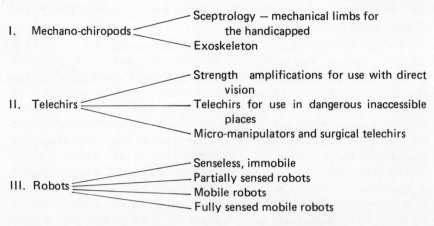

I.   Mechano-chiropods
- Sceptrology — mechanical limbs for the handicapped
- Exoskeleton

II. Telechirs
- Strength amplifications for use with direct vision
- Telechirs for use in dangerous inaccessible places
- Micro-manipulators and surgical telechirs

III. Robots
- Senseless, immobile
- Partially sensed robots
- Mobile robots
- Fully sensed mobile robots

Fig. 1.2 – Classification of systems with humanoid limbs.

The *second group* is telechirs and our knowledge about them, which is described by the word 'telechirics'. This word means 'hands at a distance' and it implies putting a great part of the trained skill of a craftsman or machine operator at the other end of a communication system so that he can do a variety of skilled operations with a skill approaching that which he would have on the spot. The human is in direct control of the telechir in real time all the time it is working. He may be quite close to the telechir in which case his visual control can be direct. Examples are the giant powered arm operated by a driver in a carriage, the mechanical remote hands used in radioactive caves and the powered mechanical remote hands used in undersea, one-atmosphere pressure vessels. It will never be possible to give the operator all the craft skill of the direct use of hand tools in his own hands and the hands will have to work more slowly. However, these disadvantages will in many cases be more than compensated by the advantages of the human operator being in a safe, comfortable place or being able to work giant arms of great strength or miniature arms of great precision. A generalised diagram of a telechir and its necessary components is shown in Fig. 8.1. The most important use of telechirs will undoubtedly be for work requiring the adaptability and the coordinated hand—eye craft skill of a man in dangerous, remote or uncomfortable situations, such as radio-active areas, sewers, mines for coal or other minerals, work on the bottom of the sea, work on explosives, putting out building fires or rescuing people from them and re-bricking of red-hot furnaces from inside. The characteristic of the machine for these purposes is that it must have a body capable of operating in the hostile environment and in general being moved about in it, carrying arms and hands and tools for the various tasks it has to perform, and a visual and tactile feedback system. The human operator is connected to it all the time it is working and thus the skill of all his three brains is available, particularly the 'emotional brain' to give the adaptability and creativity to cope with a totally unexpected situation. The 'intellectual brain' is needed to calculate the optimum solution, while the craft skills of the 'physical brain' carry it out. However, the telechir may also have a certain element of robot ability incorporated in it to save the human from repeating the more detailed actions and to ensure that nothing is forgotten. These might be a programme to carry out a routine operation, such as replacing a pick on a coal cutting machine when the human would simply carry out the programme once the broken pick had been detected by his senses working remotely. Another example might be the instruction to move to a certain place and make its own decisions as to how to avoid obstacles on the way there.

The *third group* is the robot. The word *robot* comes from Capek's play RUR, in which human actors always take the part of robots. It may be defined as a machine with a computer memory which can be programmed by a human to carry out any one of a variety of different series of complex manipulative tasks, for example the first generation, 'senseless' robots have an arm and hand so that they can be programmed to carry out and repeat a series of up to 200 successive operations such as picking up an object, moving it and putting it down, feeding it into a machine, operating the machine, taking it out and so on.

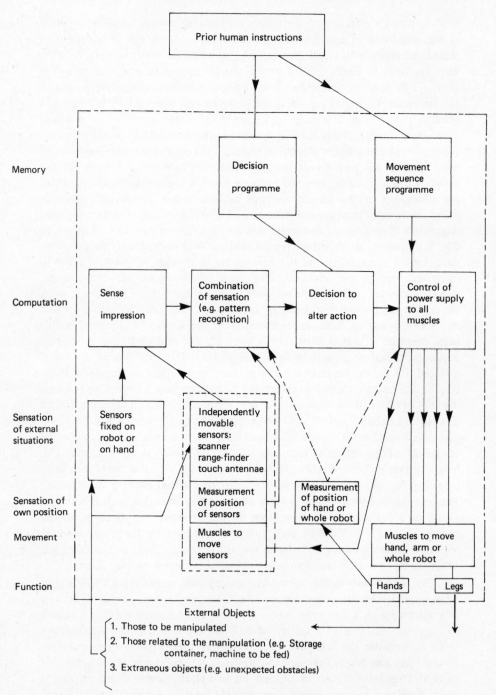

Fig. 1.3 – Block diagram of control system of generalised robot.

Once the human has instructed the robot he leaves it to repeat this series of movements hundreds of times until it breaks down or the need for that task is completed and then it can be programmed to do a different one. However, first generation, 'senseless' robots will still try to carry out their task even though the object to be picked is not there or something gets in the way.

Second generation robots are fixed in space but can adapt their arm and hand movements as a result of simple sensory observations.

Figure 1.3 is a block diagram of the control system of a generalised third generation robot which has senses such as vision and touch and can process this information by means of its computer to modify its movements in a way pre-instructed by a human. The third generation robot can have the ability to move itself about its environment and avoid obstacles.

Figure 1.4 gives a very crude and simplified survey of the probable comparison of the abilities of a man's hands and those of a robot and a telechir. Compared with a human worker a robot has certain advantages.

Attribute	Man	Robot	Telechir
Long hours of work	Attention and skill fluctuates, needs tea and meal breaks and sleep	Tireless	Succession of human operators control single telechir
Size	1.5–2.0 m	0.5–5 m	0.1–5 m
Max. strength	50 kg_f	$1T_f$	$1T_f$
Sensitivity	1 mgm_f	10 g_f	10 mgm_f
Multiple attention	Main and subsidiary	Several tasks simultaneously	1 task/human controller
Max. steady state power	¼ kW	5 kW	50 kW
Materials	Bone and muscle (no continuous rotation)	Steel, aluminium plastic	Steel, aluminium plastic
		Much stronger for given size and weight	
Access to awkward places	Moderate	Moderate	Can be made very good
Hand—eye coordination	Trainable to very high level	Poor or non-existent	Relatively clumsy
Sensor adaptability	Very good	Poor or non-existent	Can be made good
Emergency response	Limited by physical danger	Zero	Can be excellent

Fig. 1.4 – Comparison of attributes of man, robot, telechir.

(i)   It can work a 24-hr day 7 days a week except for maintenance stoppages and breakdowns. It does not need rests, breaks, sleep, meal times or any other work interruptions and in many jobs this increases output more than proportionately if it is serving a machine that needs time to warm up after a break. It does not get bored or careless when tired and accidents to it are of far less human consequence than those affecting humans.

(ii)   It can be up to 100 times stronger and more powerful and have a reach up to 10 times that of a human and can put its steel hand inside a furnace. Scale-up laws mean that a giant made of biological muscle and bone becomes slow and ponderous and eventually subject to broken bones. These restrictions do not apply to robots which can be made of steel with many times the strength/weight ratio of bone and can be powered by engines that consume chemical energy much faster than an elephant. In a study of scale-up laws in regard to flying it has been shown an air breathing bird or aeroplane engine has a power output approximately proportional to the air intake or nostril area and a robot with an internal combustion engine for power can, like a motor car engine or a jet engine, have a very large intake. Alternatively it can have a electric motor up to 50 kW. Similarly it has been shown that no biological creature the size of a man could ever fly by flapping wings because the extra muscle that would have to be carried would increase the weight by an impossible amount. The largest birds like ostriches are not capable of flying and could never be as they could not work wings large enough to lift them into the air and some of the largest vultures that can get into the air require some assistance to do so, such as rising air currents or strong ground winds.

(iii)   The robot can have a large number of arms or control a large number of legs with its computer, to act in a coordinated way. These limbs can be double-jointed or telescopic and can be capable of continuous rotation about any axis.

1.5 PHILOSOPHY OF ROBOTS AND THEIR POSSIBILITIES AND LIMITATIONS

With the development of the computer and now the extreme miniaturisation of the microprocessor, there has been a great deal of interest in, and research into artificial intelligence and the provision of hands and arms to sophisticated computers and the question is how far this work could be carried. Mechanical intelligence systems have already been used for hypnosis and for teaching systems in which the response of the machine is different according to whether or not the pupil indicates that he has learnt the previous material by giving the right answer to a multiple-choice question. In the latter case a more detailed explanation is repeated, in the former it goes on to further material. Computers have also been used for medical diagnosis. It is questionable how far computers can replace the teacher—pupil or doctor—patient human relationship.

Computers have been developed to an extraordinary degree to help humans to store and recall very complex information, to carry out speedily numerical calculations of a kind far too complex for the human brain, involving multiple choice and vast arrays of numbers, and to analyse incoming information. Many excellent textbooks are available on computers and artificial intelligence but here we are concerned with the possibility that robots should replace humans in manual activities such as domestic tasks, the use of tools, factory assembly processes and controlling and operating other machines such as coal-cutting machines.

In the first section I considered the question of how far *it is desirable* for man to develop robots, here we are concerned with the question of how far he *can* develop them: how far an artifact can replace a human worker. Enthusiasts for the subject say it is only a question of doing enough research and spending enough money to replace humans entirely in all manual activities. A great deal of research is going on to produce second and third generation robots, that is robots with senses and sensory adaptiveness and mobility and it is already becoming clear that to attempt to copy even part of the human coordinated hand/eye skill and sensory adaptiveness, which takes us at least 10 years to learn, will be very expensive indeed in terms of the use of very expensive materials. Therefore it is unlikely that humans will be replaced in most tasks requiring coordinated hand/eye skill for purely economic reasons but some special ones may justify the development of manufacturing costs. However, the question we have to look at here is whether there is an ultimate theoretical limit to the possibilities of robots.

As a result of 20 years study I have come to the conclusion that there are fundamental *laws of impotence* in the manufacture of artifacts. The first and second laws of thermodynamics are *laws of impotence* which can never be proved logically but result from the failure of many hundreds of experiments aimed at going against them. All attempts to make perpetual motion, either without any source of external energy or by taking energy from the surrounding atmosphere have always failed. As a result of this the first and second laws of thermodynamics have been stated in forms such as:

(i)   the total energy of any closed system must stay constant so that if it is moving with friction the movement must slow down;

(ii) heat cannot flow unaided from a colder to a hotter source.

As has already been mentioned previously the subjective study of one's own human system confirms the philosophical idea that a normal human being has three brains, each of which performs different functions. As a result the human being can take part in three different kinds of event.

(i)   *Purely causal events* in which he does a certain activity and it produces a predetermined result. Such activities are primarily concerned with the intellectual brain which calculates the required causality and the physical brain which performs and controls the chosen activities. An activity with which one is completely familiar such as driving a car in traffic can, of course, be done with the physical brain alone while the intellectual brain is working on some quite other problem. The emotional brain may not be involved in the activity at all except possibly to provide the energy of decision that the task is worth doing or annoyance at being overtaken by a smaller car.

(ii) *The law of accident* in which some totally unexpected external event upsets the expected causal relation, as for example a sudden breakdown of the car or a tile falling on your head as you walk along a street. In the case of the car breakdown or other emergency immediate sudden decisions are needed to avoid an accident and these are provided by the rapid response of the emotional brain to the emergency giving a direct stimulus to the physical brain to take action far more quickly than thoughts can work. If one suddenly sees a child stepping off the pavement the emotional reaction will be far greater than if it was a cat. Only afterwards is the intellectual brain brought into action to try and decide the cause of the breakdown and a plan of campaign for detailing with it.

(iii) *Freewill.* Philosophers have always argued as to whether man has freewilll or not but a study of one's own subjective inner processes shows conclusively that one does on very rare moments have freewill, that is to say moments when one can decide to take a certain action which could not have been predicted by any purely logical processes, nor from a total knowledge of the history and heredity of the person concerned. The most important examples are the flash of inspiration leading to something truly creative and original. What De Bono calls *lateral thinking,* the solution of a problem by a totally non-logical way out, also comes into this category. This again depends on the very fast-working emotional brain, the speed being indicated by the common expression a 'flash of inspiration'.

Thus in two out of these three situations the human emotional brain is essential to successful activity. I have come to the conclusion that we can build computers to carry out the purely logical processes of the intellectual brain. Similarly we can build mechanical hands and arms and legs to carry out all the activities of the human limbs and the physical brain and also to make use of sense impressions such as sight and touch in a similar way to that which the human does, but I am fully convinced

we can never construct a genuine emotional brain in a artifact.

In religious terms, if we could, it would be the god-like activity of creating a machine with an emotional brain which has feelings such as religious feelings, value judgements, conscience and morality. It would then have all the attributes and possibilities of a man and rank equally on the scale of values. Just as all the medieval attempts to create an artifical 'homunculus' failed, so I believe all attempts to create an emotional brain in a computer or robot will fail.

Isaac Asimov has formulated the three laws of robotics as:

(1) A robot must not harm a human being, nor through inaction allow one to come to harm.

(2) A robot must always obey human beings, unless that is in conflict with the first law.

(3) A robot must protect itself from harm, unless that is in conflict with the first or second laws.

These laws are not in fact laws but principles which the human designer is recommended to apply for his own protection when designing robots. He is in fact building his value judgements into the robot and this is I believe all he can do. In the film '2001' the robot computer Hal suddenly develops negative emotions and tries to kill its human passengers. If my belief is correct it could not develop negative emotions any more than positive ones but it could, of course, kill them by coming under the law of accident, that is it could break down or go wrong.

I believe that man's inability to create an emotional brain in an artifact will be found to lead to *two laws of impotence* for robots, the first of these may be called the 'sophistication postulate'.

Law 1. A robot cannot be built to do any task more sophisticated or organised or unexpected than those in areas which its designer foresaw and for which it is programmed. A robot can be much more patient, tireless and accurate than a human in doing precisely that task for which it has been programmed but if it deviates from its instructions or is faced with a situation for which it has not been instructed the results will be random and accidental and not an improvement on the situation.

Heginbotham[†] has defined the ultimate robot possibility as creating order from disorder but he points out that the initial disorder must not be too great. For example in the case of an assembly task the components to be assembled must be available in a reasonably regular fashion and the task it has to do must be pre-designed. Another example: if it was told to sort objects into a series of categories with a compartment for each category it could not come to the conclusion after a while that an extra category is needed to produce a more ordered situation. Similarly, while a robot can be programmed to play a game of chess by using its rapidity of logic to work out the consequences of various alternative moves, it is unlikely it would ever have the emotional feeling for the strength or weakness of a move which is acquired by a grand master and it would certainly not be able to say that it was bored with chess and invent a new game. There are three corollaries to this law.

[†] W. B. Heginbothm, ECE Seminar on Industrial Robots, Teknologisk Institut Copenhagen, 1977.

Corollary 1. If we were to design a self-repairing robot it could only repair those breakdowns which were envisaged by the designer and, of course, just as with humans, certain breakdowns, such as a breakdown of the main power system, would require external assistance. In the case of unexpected breakdowns the intelligence of a human controller would be essential to diagnose and correct them. It is a law of engineering that the designer can never envisage *all* the possible ways in which a complex machine will break down and thus a robot factory must necessarily have considerable supervision and maintenance.

Corollary 2. If we able to construct a robot which could itself carry out the operations of constructing a duplicate of itself the succession would inevitably degenerate to uselessness in two of three generations because of the entry of unexpected faults into the system.

Corollary 3. A robot will never be able to cope with a totally unexpected situation of any kind.

Law 2. An artificial intelligence (robot or computer) can never have true human emotions. We can, of course, make a robot that will *simulate* the external expression of human negative emotions, such as anger, by building into it a tape recorder on which a skilled actor records the appropriate sounds and words. Emotions are necessarily personal and subjective and there is no doubt that the higher animals have them. One has only to experience the affection of a dog or a horse for its human master. This is why we have an emotional feeling for an affectionate animal, which is much closer in quality to that which we have for people than to that which we have for artifacts such as motor car or a robot.

Corollary 4. A robot can never have free will and do something really creative. A robot or computer could be programmed, for example, to ring the changes on a musical formula but it could not work out an original musical composition which would give pleasure and real feeling to people.

Corollary 5. A robot can never make its own value judgements. For example a robot could not make a 'hierarchy of values' such as that of motor car, dog, man.

Corollary 6. A robot cannot be self-motivated. It cannot decide to make a big effort to do something different or to look after its own health better.

1.6 THE POSSIBILITIES AND LIMITATIONS OF TELECHIRS

The telechir can have all the advantages of robots listed above, especially if it has its own computer or microprocessor to enable it to be programmed to carry out certain tasks, but it is not subject to the limitations of robots because a skilled and trained adaptive human is in control who will not panic. It has three considerable advantages over the situation in which the man himself is doing the task in a dangerous place.

(i)   It can be made very much more robust for working in a hostile environment, e.g. made of steel so that it can withstand rockfall; it need not breath air and it can stand much higher temperatures than a man.

(ii)   As the man's physical body is not in danger he can think calmly of the most appropriate action in an emergency.

(iii) It can have a means of locomotion and arms designed for the hostile environment, for example to work in limited and awkard positions.

The following list contains some of the hostile environments for which telechirs are being developed.

(a)   Work in space. Here the ability to work in a total vacuum and to withstand cosmic radiation and meteoroid impacts is essential.

(b)   Radioactive situations. The first developments of telechirs were made more than 20 years ago when there was a proposal for an aircraft operating with a nuclear reactor for power. This reactor would have been shielded only on the side of the body of the aircraft so that if there had been a crash the whole area would have become radioactive, or on landing no-one could approach except from the rear. Since then much work has been done on telechirs for handling radioactive materials in connection with nuclear power and explosives.

(c)   Underwater. As it becomes necessary to drill for oil at increasing depths it becomes more and more expensive to have platforms standing on legs or to maintain the position of a floating platform sufficiently accurately and the work of divers is becoming increasingly dangerous and limited. At present work is being done to enable humans to do the work done by divers while situated in air at 1 atm. pressure. Ultimately an oil well could be drilled and connected to a pipeline on the seabed by means of telechirs operated by men in ships or on shore.

(d)   Military purposes. Much of the work on telechirs, in the past, has been for these purposes. Simple telechirs are already being used to enable terrorists' bombs to be defused or exploded without damage to humans.

(e)   Mining for coal and other minerals. Here the advantages of not having to ventilate the mine, of being able to go to depths at which the temperature is too high for humans and of being able to work thin seams economically and seams far under the sea make this development potentially capable of giving the whole of mankind enough energy and enough metal ores to last for more than a hundred years.

(f)   Miscellaneous other purposes, such as fire-fighting, work in industrial furnaces, and rescue of humans in a disaster situation.

Many of the problems of telechirs can be solved simultaneously for all these applications, so that the control and information transfer for telechirs provides a field of research of immense human value; because once these problems have been solved telechirs could be mass-produced in different sizes and with different bodies. There will probably always be certain limitations in regard to the hand/eye coordinated skill, and the rapidity of movement and the communication of the details of the situation to the human operator but all these can be greatly reduced by research and development applying known principles.

1.7 LIMBS FOR THE DISABLED

Table 1.1 gives a 1971 estimate of the disabled people in the USA, which had a total population of 150 million. This table was estimated by Dr James Reswick,

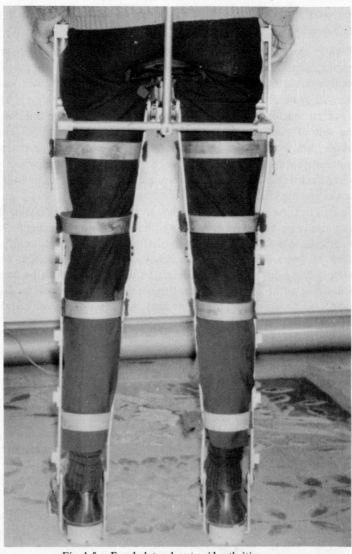

Fig. 1.5 — Exoskeleton legs to aid arthritics.

Table 1.1

Estimate of disabled persons in the United States (1971).[a]

Disease category	Total number of patients	Total no. in Rehabilitation centers	Number of new patients/year	Percent rehabilitable	Number that could be helped with current technology
Stroke	2,000,000	250,000	500,000	60	400,000
Cerebral Palsy[b]	550,000	?	50,000	45	500,000
Multiple Sclerosis	500,000	?	50,000	?	50,000
Spinal Injuries*	100,000	1,000	10,000	95 (80 to jobs)	100,000
Amputees	350,000	?	?	100	600,000
Rheumatoid Arthritis	13,000,000	?	?	?	1,000,000
Totals	19,500,000	—	—	—	3,000,000

*Liberty Mutual estimates total cost for each quadruplegic patient, ranges from $250,000 to $350,000; direct medical treatment costs range from $25,000 to $35,000 per patient, the remaining costs cover such things as workman's compensation, extended care, etc.

[a]Estimates supplied by Dr James Reswick, Rancho Los Amigos Hospital, assembled from sources ranging from 'hard' to 'soft'. Orders of magnitude appear correct.
[b]1956 figures.

Rancho Los Amigos Hospital, and was given in a paper 'Impacts of Telemation on Modern Society' by A. D. Alexander III, *First CISM IFToMM Symposium on Theory and Practice of Manipulators,* Vol. II. p. 12, (Springer-Verlag, 1974). Injuries to the neck can cause paralysis of the arms & legs. With the exception of diabetes patients, mechanical limbs could be of value to patients suffering from all the other disabilities. Figure 1.5 shows a pair of exoskeleton legs which enable a person with arthritis causing pain due to the pressure on the leg joints, to stand and walk with his own muscular energy while the greater part of the weight is taken off the painful joints and taken through the exoskeleton from the bicycle saddle to the metal pads under the feet. The table does show how many more people suffer from rheumatoid arthritis than from all the other disablements added together.

This is an illustration of the usual medical dilemma of where to put the bulk of research energy; either on to the rare but spectacular treatments (e.g. heart transplants) or on to the study of ailments from which millions of people suffer but which are much more difficult to alleviate except by pain killing drugs. In Chapters 5 and 6 dealing with mechanical arms and legs respectively I shall give an account of work of aiding handicapped in these two areas.

CHAPTER 2

Man as a machine and as a maker of machines

2.1 THE HUMAN BRAINS

There are three ways of studying the human brains. The first method of study is the subjective way of observing oneself think and act. It has already been suggested in Chapter 1 that this subjective study has led ancient civilisations that have studied the inner working of a man in detail to the conclusion that he has three brains, intellectual, emotional and physical. Of these the intellectual brain is far the slowest, working with logic and abstractions from reality (e.g. word-concepts) but the intellect is the seat of normal consciousness as we only become conscious of our emotions and our actions on certain special occasions, for example when we are learning to drive a car, or when we are very upset emotionally. These self-observations have also led to the realisation that most of the time a human being is no more than a machine and that even the intellectual processes are automatic responses of an associative sequence to external or internal stimuli, such as bodily discomfort due to having over-eaten, or a nagging worry about something which might happen the next day (e.g. having to catch an early train). An obvious example of the mechanicalness of thought processes due to external stimuli is when one's eye becomes caught by some random observation of printed words such as a newspaper headline or an advertisement and an unintended associative chain of thoughts is started as a result. However, as has been said, there are rare occasions in which human beings act with free will, that is to say in a manner which could not have been predicted by an outside observer and which is based on the conscious decision to go against mechanicalness for a clearly thought-out purpose. Examples are (1) the decision to leave a well-paid job and go to a less interesting and badly paid one because one felt ones employers were doing something against ones conscience; (2) the decision of someone to leave the ordinary world and become a monk because he felt that he could help the world better by doing so. As we have already seen it is this free will of a human being which distinguishes him from the most elaborate artifact which could ever theoretically be built.

The second method of studying the human brain is the method of the experimental psychologist in the laboratory. Much of the work in such laboratories is done on animals such as rats, which are not capable of doing the kind of complex organised work activities which we want robots or computers to do, and can therefore only teach us about the working of the physical or body brain and possibly also about the effects of situations such as overcrowding, upon the body. However laboratory work is done on human attention, ability to do skilled movements when tired, speed of response and speed of movement in various situations, such as a coalminer crawling between pit props in longwall mining.

Work giving considerably more insight into the brain action of humans is based on putting electrodes into the brains of monkeys and then getting them to carry out manipulative tasks. Sherington[†] studied muscle movements in animals whose spinal cord was severed so that the motor neurons were disconnected from the brain, and found that special rhythm generators could still produce rhythmical movements of walking and scratching. He introduced the term 'proprioception' to describe the sensory inputs fed back from the actual movements of the muscles — sensing: (1) elongation or position (2) tension or force.

The third method of studying the human brain is the method of the neurophysiologist and neuroanatomist who cuts up the brains of monkeys and stains them in order to study their structure and carries out post mortems on humans. He can also put micro-electrical probes into living brains (of animals) to observe their living behavious and external electrodes on humans to observe the 'brain waves'. This is to rega.d the brain as a mechanism like a very sophisticated computer and try to work out the functions of the components and the 'wiring diagrams'. The results of many years of this work can be simplified and approximated as follows.

The brain of a normal human being weighs about 1.5 kg and contains between 10^{10} and 10^{12} neurons, say about 10^{11}, and about 10^{14} synapses.[‡] The two cerebral hemispheres which occupy the top half of the skull are covered with the cerebral cortex which is a folded plate of neural tissue of total area about 0.25 m^2 and of thickness 2 mm. This contains about 10^5 neurons per sq. mm. The neuron is the fundamental unit of reception for external stimuli and of the thought process. It is roughly analogous to the on—off switch element of a computer. Each neuron (see Fig. 2.1) consists of a cell body of diameter 5—100 μm. It receives incoming signals, combines and integrates them into an average which causes it to emit an outgoing signal. The outgoing signal is transmitted along a main conductor called the axon which is somewhat analogous to the trunk

[†]See 'Brain mechanisms of movement', E. V. Evarts, *The Brain*, p. 98 (Scientific American Book) 1979.

[‡]A number of the same order as the number of stars in our galaxy.

of a tree. The axon is 10–20 μm in diameter and 0.1–1 m in length. It has branches coming off it so that it can transmit its outgoing signals to other neurons and to terminals. The incoming signals come in to the neurons through dendrites which are like the roots of a tree but here we come to the real and almost unbelievable complexity of the human brain: *each neuron is fed by signals from hundreds or thousands of other neurons.* This means that the brain does not work by a linear sequential programme but is equivalent to hundreds of thousands of complex circuits in series or parallel, richly cross-linked.

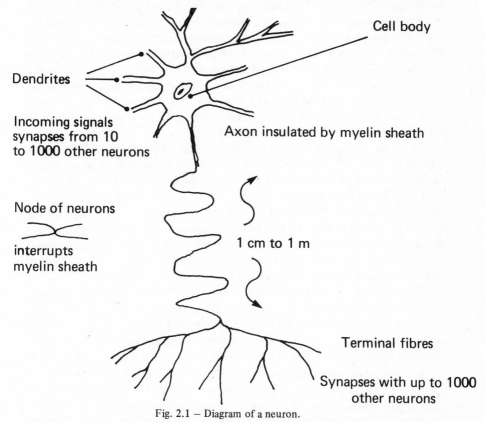

Cell body

Dendrites

Incoming signals synapses from 10 to 1000 other neurons

Axon insulated by myelin sheath

Node of neurons

interrupts myelin sheath

1 cm to 1 m

Terminal fibres

Synapses with up to 1000 other neurons

Fig. 2.1 – Diagram of a neuron.

However, the complex interconnections among the neurons are highly structured and specific. To construct a circuit of such incredible complexity by human means would appear to be beyond the bounds of possibility. A baby is born with virtually all the neurons it will have for its whole life so that these have been constructed by the growth process at a rate averaging 250,000 per minute throughout the nine months before birth. These cells are packed together in a densely intertwined thicket. They can only be distinguished under the microscope if less than 5% of them are selectively stained.

The signalling between neurons is both electrical and chemical. The neuron generates an electrical signal along the axon but the contact between the axon terminals and the dendrites of other cells is by a molecular flow across the contact junction which is called a *synapse* and there are 10^{14} synapses in the brain. Each neuron has 10^3-10^4 synapses and can receive information from 1000 other neurons. The fuel for the neurons is glucose which is brought to them by the bloodstream. The neurons act as a kind of fuel cell whereby they maintain themselves with a negative potential of 70 mV inside, compared with the outside surroundings. The electrical signals only operate for short distances, of the order of 1 millimetre and they operate by changing the potential differences along the axon. For longer distances, such as the muscular control nerves to the feet or the sensation nerves from the feet, the chemical signal is used within the axon and the signal transmission time is of the order of milliseconds, very much slower than the communications in a computer which are small fractions of microseconds. The axon is insulated by a sheath of myelin but this is interrupted roughly every millimetre by narrow gaps (nodes of Ranvier) where the signal is passed on chemically by the following process.

Normally the concentration of Na/K in the exterior fluid is 10/1 while inside it is 1/10. This is concentration difference is maintained by an ion pump. The operation of the chemical signal within the axon is that the polarisation voltage of the membrane is reduced from -70 down to -50 mV and this removes the normal barriers to the flow of sodium and potassium ions through the membrane wall; the resulting flow reverses the potential to $+50$ mV producing a strong signal which is transmitted along the nerve fibre, thus the signal cannot have more than an on—off information content and it is *the number of impulses per second* which transmits the analogue message along the nerve fibre.

W. S. McCulloch (*Embodiments of Mind*, MIT Press, 1965) has described the human nervous system in a way which enables it to be compared to the control system of a robot and Fig. 2.2 is a simplified version of a diagram based on his work (Assembly of computers to command and control a robot, L. L. Sutro and W. L. Kilner, paper to Spring Joint Computer Conference, Boston, May 1969).

Regarding the human being purely as a system for manipulating objects in the outside world (which is all we expect of a robot) we can neglect all the marvellously complex internal self-regulating systems for controlling temperature, breathing, chemical composition and biological constituents of the blood, blood flow carrying oxygen and fuel to the muscle and brain cells, hormones and glandular secretions. The self-constructing and self-repairing systems of the human body are also outside our engineering potentiality, as is even more, all the inner activity of a human being: hopes and fears and strivings for self-development. We can thus represent a human being in the aspect of an object manipulator with craft skills but without the creative originality of an artist as in Fig. 2.3. In this diagram no attempt is made to plan the elements in their correct relative position.

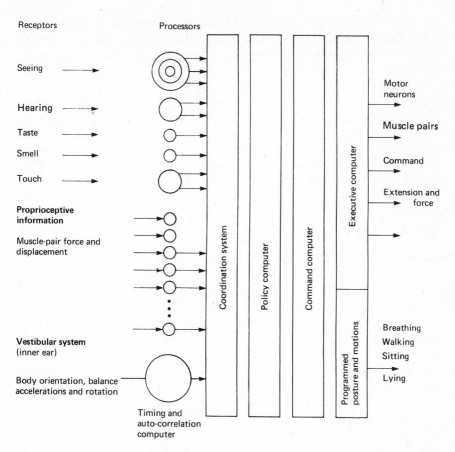

Fig. 2.2 — The human as a manipulative system.

Information comes into the system from three groups of sensors. First we have the four senses with specifically located receiving areas, sight, taste, smell and sound. Each of these senses has its own processors which store information from past experiences so that it can be compared as when we recognise a smell or a piece of music. The senses most important in the manipulative connection are sight and touch; sight will be discussed in a separate section.

Touch, and sensations of heat and pain can come from any area of the body and are processed via the spine like the second form of information as to the force exerted by the joint and its rotational position. All this information is processed to give an overall 'muscular' model of the human body in its interaction with the outside world and particularly with the manual task or bodily movement being undertaken.

The third group of sensors is the vestibular system of the inner ear which senses body orientation in relation to gravity, balance, accelerations and rotations. The vestibular system for example makes us giddy with excessive rotation.

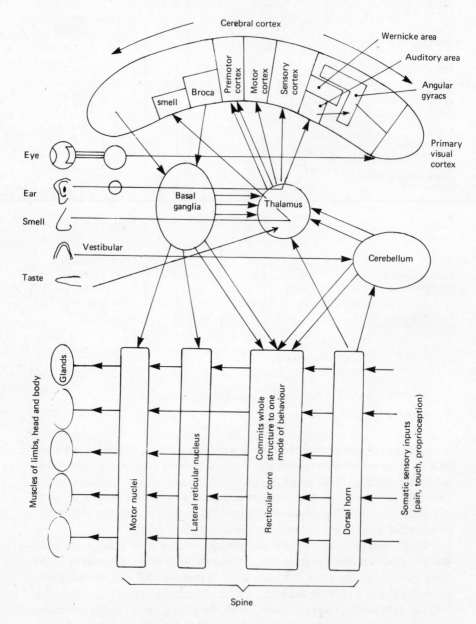

Fig. 2.3 – Simplified diagram of central nervous system.

All these three sensory systems provide information which is grouped in sensory ganglia (clusters of neurons) located in the spinal chord or flanking the brain. In Fig. 2.3 the components of the human nervous system are shown very diagrammatically in their relative positions and with the most important connections. The cerebral cortex is the main computer which completes the processing of incoming information from the eyes, ears and nose and the nerves of the whole body. It carries the stored memory of acquired skills such as language (Broca's and Wernicke's areas) the use of tools (motor cortex) and interpretation of visual pictures (primary visual cortex). It makes long-range policy and carries out many other functions including the emotional activity (limbic forebrain) and memory of abstract ideas, guiding principles, morality. Little is known about these processes and these areas of the cortex and it is probable that the method of self-observation is the only one deep enough to study these subtle aspects. What is quite certain is that we shall never be able to build them into an artifact – we cannot even hope for a wiring diagram of 1 mm^3 of the human brain and the way all its neurons are are firing (F. H. C. Crick, 'Thinking about the Brain', *The Brain*, Scientific American Book, p. 133), nor can we expect to construct a computer with 10^{11} components (neurons) and 10^{15} synapses or contacts between them; each of these synapses can be strengthened or weakened in the learning process.

The somatic sensory cortex and the motor cortex are like bands that run across the top of the head from ear to ear – the motor cortex is the front band.

A nerve impulse (e.g. a pin prick) is transmitted via the spinal column and the thalamus to the somatic sensory cortex in such a way that one is aware of the exact position of the prick. The fingertip is especially sensitive and can separate two simultaneous pricks 2–3 mm apart. Similarly the signals from the eye, nose or ears after processing in their special areas of the cortex pass to a region of the sensory cortex. Parts of the body which are extra-sensitive occupy large areas while the trunk and legs occupy very small areas.

Information received in the sensory cortex goes through a very little-understood processing in the other areas of the cortex and this results in a plan of action which is transmitted to the motor cortical area. This has areas corresponding to all the muscles that we can consciously control, including the neck, brow and swallowing mechanism, which do not have separate sensory areas. The sensory region, however, has sections for the intra-abdominal (stomach ache), pharynx (sore throat) nose, forearm and genitals which are sensitive but not under conscious muscular control.

The somatic sensory inputs from all the muscles and skin of the body (pain, touch and proprioceptive) are transmitted via the dorsal horn, which is the sensory elaboration of the spinal cord, to the cerebellum and the thalamus. The cerebellum computes the termination of movements, so that we have extraordinary skill in moving for example the hand rapidly to the exact place decided upon or regulating foot movements in running on irregular ground.

It is also an internal clock and autocorrelator and receives inputs from the vestibular system to detect tilt and acceleration of the head and from its inform-ation from the touch and proprioceptive sensors detects posture and the nature and position of what is being touched by hand or foot.

The thalamus is the principal switching structure which receives processed information from the senses, the dorsal horn and the reticular core and passes them on to the appropriate areas of the cortex such as the pre-motor cortex, the somatic sensory cortex, the language areas and the primary visual cortex.

After the highly complex, and not understood, processing in the cerebral cortex the motor cortex sends the control signals for the body muscles via the basal ganglia to the reticular core of the spine which commits the whole body to one mode of behavious and commands the appropriate muscular movements.

The basal ganglia contains the store of innate or learned total action patterns such as feeding, walking or throwing a ball. It acts on all the other computers and they report to it. The lateral reticular nucleus is another spinal computer which controls breathing and other routine bodily functions (digestion, heartbeat, glandular activity).

In more primitive animals the cerebral cortex with its 'conscious' control of movement through the motor cortex does not exist. The animal lives and moves by means of the basal ganglia, the cerebellum, the main body computer, the reticular core of the spine and the other three spinal mechanisms shown in Fig. 2.3.

Approaching the problem of understanding the brain processes from the self-observation direction, one can see processes such as abstract thought and metaphysical yearnings which distinguish the human from the animal.

Moreover the human brain contains many highly sophisticated specialist abilities and one of the things which distinguishes it totally even from monkey brains is its capability of learning a variety of more specialist activities.

Some of these special activities which have to be learned are the following.

(1) Language. There are two limited areas in the cerebral cortex which have been found to contain the linguistic skills because people in whom these areas are damaged lose their ability with language; curiously these are mainly on the left side of the brain.

(2) Music. Most people can learn and remember a number of melodies.

(3) The ability to draw simple figures or pictures which are related to real objects and can be used to communicate.

(4) Ability to do simple but accurate mental arithmetic. This is especially valuable if it is connected with the physical experience of size or quantity or the emotional feeling of value.

(5) The recognition of faces. This is a very highly developed skill, useful for human beings in a social context, and has a special area on both sides of the brain underneath the cerebral cortex.

Looking at the brain as an energy converter it weighs only 2% of the body weight and yet uses 20% of the oxygen consumed during the quiescent state of the body. This, however, only corresponds to some 20 watts/day and night. Whereas the muscles can use chemical energy of food in various forms such as sugar, fats and amino-acids, the brain can only use its energy in the form of glucose in the blood. Thus neurons which are active in the brain at any given moment take up glucose more rapidly than the quiescent ones, requiring this energy for their firing activity.

2.2 THE CEREBELLUM

The cerebellum controls the coordinated behaviour of the entire human moving system. It regulates movements, muscle tone and balance and it is the watershed between sensory and motor processes. It has been compared to a matrix-type computer in which a whole row of incoming signals are compared and coordinated and a corresponding series of commands goes out to groups of muscles from the columns of the matrix. It receives input signals from: (1) the cerebral cortex of the movements required; and (2) the vestibular system which gives information on the tilts and acceleration of the head necessary for static and dynamic balance; (3) proprioceptors, the sense cells of the skin and muscles and the joints such as the overall posture and the nature and position of what is being held. The cerebral cortex will have worked out the necessary modifications to the general instruction for walking or running in a given direction, resulting from visual estimates of the terrain in front, but the cerebellum instructs the muscles via the spinal cord to take the necessary altered action. It computes the termination of a movement, such as reaching to touch an object and computes the necessary muscular reaction to catching a weight thrown to the person. This is why a person drops a small sphere of uranium when it is thrown to them, because they estimate its density as that of steel. Similarly a wrong estimate of ground friction or hardness can lead to falling over.

There are two kinds of synapses in the cerebellum, the non-modifiable ones which deal with reflex activities and the modifiable ones which enable a person to learn complex movements like walking and balancing or picking up an object. It has been estimated that the number of control elements in the most sophisticated artifact is less than one part in 10,000, compared with the human system.

2.3 MOVEMENT[†]

The human body is capable of precisely controlled movements ranging from swinging a sledgehammer to land accurately on a wedge for splitting wood to the movements of a surgeon in microsurgery, for example on the eye, working under a low-powered microscope. Human movements can be divided into three classes.

[†]See 'Brain mechanisms of movement', E. V. Evarts, *The Brain,* p. 98, Scientific American Book, 1979.

(1) Totally reflex actions in which there is no consciousness at all. These involve only muscles and the motor neurons of the spinal cord. For example, if the connection between the cortex and the spinal column is broken breathing is still possible but the muscles are not any longer useful for speech or song. Scratching in response to an itch and the well-known knee reflex are other examples. Even automatic walking in a dog is possible after the connection between brain and spinal colum is severed.

(2) Movements that involve a mixture of volitional control and processes fully learned but unconscious, such as two-legged walking with its sophisticated balance system. These are operated by spinal rhythm generators. These have automatic inputs such as visual estimation of the state of the ground on which the foot is to be placed but these pass through the sensory and motor cortices and the walker has cerebral volitional control to decide to walk. The consciousness in the cerebral cortex decides the goal, such as when a skilled marksman concentrates all attention on the target while his body holds the gun steady by compensating for all different movements; similarly a singer may concentrate on the note to produce and a tightrope walker on the use of the pole for balancing.

(3) Completely voluntary movements in which cerebral conscious attention has to control all the stages. These apparently involve the basal ganglia of the brain and the cerebellum. An example might be when one steers a car for the very first time before the reflex is set up by experience; or when one is giving full mental attention to a craft activity such as woodcarving. It is easy to observe how much clumsier and slower the movements are when one has to control them by intellectual attention.

Any movement such as walking requires the fully coordinated contraction and relaxation of many muscles. There are two kinds of muscles, one typified by the muscles that control the movements of the eye which use comparatively little force but control with great speed and precision to a few minutes of arc. These muscles have what is called a high innervation ratio, that is the ratio of the number of motor neurons[†] to the number of muscle cells controlled by these neurons. For the eye this ratio is about 1 to 3, one motor neuron is needed to control three muscle cells.

The other kind of muscle is typified by the great muscles of the arm and legs where the innervation ratio is 1 to several hundreds, that is to say the output from a single motor neuron controls several hundred muscle cells and produces a coarse twitch as the minimum movement.

Each muscle is an inter-mixture of two kinds of motor units, the slow twitch motor units which have a great resistance to fatigue but give small tension, and the fast twitch motor units which give a very large peak muscle tension but fatigue

---

[†] A motor neuron is a neuron with axon which terminates on the outer membrane of the muscle cell; these run from the spine to the muscles.

rapidly. The fast twitch motor units are controlled by axons of very large diameter and rapid nerve impulse conduction so that a large signal can be given them.

The muscle tension is regulated by: (a) control of the number of motor units recruited to act; (b) control of the firing frequency of the motor units which have been recruited. The slow twitch units are used first and the fast twitch ones last when a big output is required. In the calf the tension of the largest motor units is some 200 times that of the smallest units.

The impulse from the end of the axon of the motor neuron is transmitted through the membrane of the muscle cell by the emission of the chemical acetylcholine. This transmission across the membrane is blocked by the drug curare and it has been found that people given a dose of this drug can continue to think inside themselves but cannot give any outward sign of consciousness because they can move none of their muscles.

Precise small movements are controlled by negative feedback to the motor cortex so that its output is automatically regulated. One source of these signals is the sensory area of the cerebral cortex which is directly behind the motor cortex. An example of this automatic regulation is the closed loop response mode that enables a person to remain upright by stepping forward when pushed suddenly from behind. The sensory feedback from the movement itself is proprioception. There are two elements to this sensory feedback both giving negative feedback for control, one is the observation of the elongation of the muscle and the other is that of the tension force in the muscle. The automatic response with negative feedback leads to an increase in the muscle length being detected and being controlled by resulting contraction. Similarly an increased tension being sensed leads to a reduction of the tension. The example given is that of a man who has decided to hold his arm out sideways horizontally. As he gets tired his arm droops involuntarily and this causes the muscles holding it up to lengthen and to reduce its tension. The lengthening excites the motor neurons of the muscle to cause it to contract while the reduction of tension removes the inhibition of the muscle so that it increases its tension by contracting. If on the other hand a man deliberately picks up a heavy weight then the increase in tension absorbed by the tension receptors is voluntarily controlled to cope with the required weight.

In addition to the normal alpha motor neurons that control the main muscle fibres there are another type called gamma motor neurons which act on the special small muscle fibres that regulate the sensitivity of the length receptors and optimise their performance.

2.4 WALKING

The human two-legged walking and balancing system is an extremely sophisticated control system and each successive movement requires the coordination of several muscles. Figure 2.4 is a simplified diagram of the bone and muscle system of the human leg and foot. The dynamic forces necessary to propel the body forward are taken by the leg through the hip joint and this, plus the dynamic

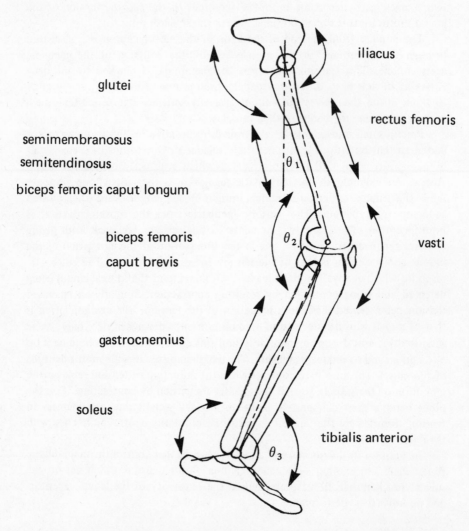

Fig. 2.4 – Simplified diagram of bone and muscle system of human leg and foot.

and static forces necessary to support the leg, adds up to the ground reaction which consists of a vertical weight support component and a forward force

walking (*3rd CiSM–IFToMM Symposium,* p. 147, 'Coordination and bio-mechanics of muscles in human locomotion', C. Frigo and A. Pedotti) which used a force plate under the foot and photographs of the position of four markers on the body and the three important joints, namely the hip, knee and ankle has made it possible to calculate the forces and moments on these three important joints at all stages of the walking cycle, that is the period when one leg is station-ary on the ground and the period when it is coming forward for the next step. A complex computer calculation knowing the ground reactions, the kinematics of the movements, the body parameters and the skeletal and muscular parameters makes it possible to calculate the detailed forces and the external energy input to each of the eleven important muscles.

The muscle energy input can go through to external work output when it is contracting against the external forces to do work, but it also absorbs chemical energy when it is *absorbing* external work by stopping a heavy moving body. This means that the muscular system does not have regenerative braking possibil-ities as a spring or pendulum would have because its energy conversion process is an irreversible chemical reaction. The muscle also wastes energy by inefficiency when it is doing external work and by tension when it is not changing in length at all, e.g. supporting a heavy weight. Thus measurements of the external work input by the consumption of chemical energy in the muscle are not clearly related to the external work output and this is why the direct measurement of external work output by these authors is necessary.

The joints of the human leg are not in fact pin joints but rather complicated ball and socket rolling movements as is found when it is necessary to produce a mechanical movement corresponding closely to the kneee movement. Never-theless, for this analysis it is possible to treat the human leg as if it were three rigid links joined by pinned joints at hip, base and ankle. The toe-bending is also ignored. The results show the complexity of the walking system compared with a simple mechanical leg with only two powered movements (leg swing and foot raise). The muscles which produce the peak rate of energy consumption to power walking are the iliacus which swings the leg forward, the glutei which pulls the leg backward and the soleus which pushes the ground backward by the foot. On the other hand the muscles which absorb energy fastest to produce a springing action are the hamstring muscle (semi-membranosus) and the knee-bending caput brevis. When, however, one studies the overall power produced by the muscle, per unit volume, counting absorption as positive, then the soleus muscle that pushes the big toe downward has by far the highest, followed by the gastro-cnemius muscle which flexes the knee and the ankle. During the period when the leg is stationary and carrying the weight the iliacus muscle that flexes the hip has the greatest power output and the semi-membranosus (hamstring) muscle which extends the hip and flexes the knee, is the greatest power absorber. The ham-strings also absorb energy in the last part of the leg swing phase and produce it just after the heel strike. The vasti muscles have their main action in the yielding

provided by the friction of the ground to propel the body. A detailed study of phase when the body weight is loading the limb after the heel strike and this acts as a shock absorber mechanism. Similarly the tibialis anterior muscle at the front of the foot begins its activity a little before the heel strike and goes on by preventing the foot drop and absorbing energy, because the muscular system is not capable of regenerating energy by reversing the chemical reaction.

2.5 VISION[†]

Vision (see Fig. 2.5) is clearly an astonishingly complex process. We learn fairly early in life to correct what we see so that we recognise an object as identical, regardless of wide variations in the viewing angle, the distance and the illumination (magnitude and direction, causing shadows) and to match the object seen, with past experience of objects which may have been seen under very different conditions. It takes a child quite a long time to learn to focus the eye and make these matching corrections and much longer to categorise objects seen: a faculty unique to humans. The retina or screen of the eye is already a complex mechanism,

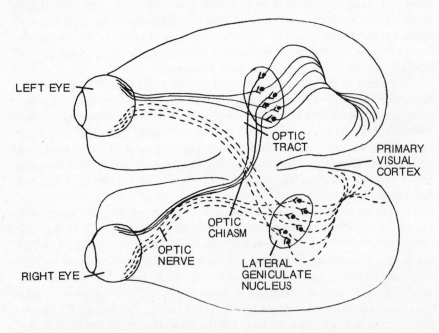

Fig. 2.5 – Mechanism of human vision (reproduced with permission, from D. H. Hubel and T. N. Wiesel, *The Brain*, Scientific American Book, 1979).

[†]Brain mechanisms of vision, D. H. Hubel and T. N. Wiesel, *The Brain*, p. 98, Scientific American Book, 1979.

the colour- and light-sensitive rods and cones pass their information through two to four synapses to four other types of retinal cells and the information is grouped in the ganglion cells, the axons of which form the optic nerve. There are about 1 million of these axons from the two eyes.

As a result of observations on individuals who have had accidental injuries leading to visual defects, it has been possible to map the critical areas concerned. Most of the sensory motor areas are found to contain systematic two-dimensional maps of the world information received. There is a curious arrangement that the axons of the cells dealing with the left half of both eyes feed to the brain region dealing with vision in the right side of the brain and conversely with the two right halves. These regions which accept the signals from the ganglia are known as the lateral geniculate nuclei and these are situated deep in the central part of the brain. After considerable processing in this region they then proceed to the two surface regions, the primary visual cortices. These consist of some six layers of cells spread uniformly in each layer, interspersed by sparsely populated layers. It is believed that in these regions the activation of the cell corresponds to a map of what was received by the retina. There is a feedback from the primary visual cortex to the lateral geniculate nuclei whose function is not known and a feed-forward to neighbouring cortical areas and several targets deep in the brain.

By using microelectrodes on single nerve fibres of monkeys and then varying the light source of the retina and observing the firing of the nuclei it has been found possible to observe what is the best stimulus for firing a particular cell. In this way it has been found that the retinal ganglion and the geniculate cells respond best to a circular spot of light of particular size in a particular place surrounded by a dark boundary. In other words they compare the light on themselves with the surroundings and look for sharp differences. The eye is very much more sensitive in the central area than on the extreme edges.

The visual cortex re-arranges the information to respond to specifically orientated line elements so that they can map the boundary of an object with a sharp outline. Thus it is possible to detect the orientation of a line switching rapidly across the field of vision. The progression through the six layers of the cortex is some kind of hierarchy in which the information gathering cells feed more complex ones, but the ultimate synthesis is not known.

2.6 CONCLUSIONS

One can conclude that not only is the human brain very much more complicated than anything we can hope to construct artificially in regard to the number of neurons and their fantastically complicated interaction, but also that there are organisations and parts of the brain which we do not understand at all, which carry out complicated processing of the information which ultimately enable our consciousness to construct quite a good understanding of what exists in the changing three-dimensional world outside us. The problems of human motivation, consciousness, will-power and ability for self-development are much further away.

CHAPTER 3

The Man/Machine Interface

3.1 THE MAN/MACHINE RELATIONSHIP

Any form of tool or complex machine used by a man to aid his achievement of a purpose forms a man/machine system with him. The purpose may be to get himself or goods from one place to another (as in the case of a car or lorry and its driver); it may be to communicate (as in the case of a telephone system); or it may be to manufacture some useful object or to mine coal out of the ground. In any case the humans provide the objective, design and build the machines, monitor, superivse and maintain them.

Although not perfectly accurate, it is useful to describe a hierarchy of man/machine systems as the machine becomes more complex and relieves man of more and more of the mechanical tasks.

Level 1 is the simple machine in which a man provides the power as well as the controlling skill; some examples are the woodman's axe, the carpenter's saw, mallet and chisel, and the hand-pulled or -pushed cart, trolley, or lawn mower. This is illustrated by Fig. 3.1. The tool or machine can be regarded as an extension of the man's body since he grasps it firmly and guides its motion directly with his own muscular effort. When using a mallet and chisel, as in woodcarving, one hand controls the position and three-dimensional orientation of the chisel while the other controls very accurately the blows of the mallet (direction, momentum, frequency).

Fig. 3.1 – Hand tools.

Level 2 is the powered machine or tool. The power may come from an animal as in the case of a horse-drawn cart or bullock-operated plough, from wind as for the sailing ship and the windmill, from the internal combustion engine as for the car or from an electric motor as for a power drill. Here the man is fully responsiple for controlling the system but he requires devices such as reins, switches, pedals and steering wheels to enable him to control the power system, as illustrated in Fig. 3.2. He may be using both hands and both feet on four separate operations, as when changing gear on a car.

Fig. 3.2 – Powered machines or tools.

Level 3 is simple automation. The single-process machine or tool with built-in auto-control enabling it to carry out this process can be left to do its one process whenever it is fed with the raw materials by the operator. It will always have its own source of power. Examples are the automatic capstan lathe in use early in this century where a drum beneath the lathe carried a set of control cams which mechanically worked the series of capstan operations. A skilled mechanic had to set up the cams for each new job. A more recent example is the numerically controlled machine tool in which a punched tape or other memory system is prepared by the human, or even by a computer, from the design and enables the machine to carry out any complex series of operations within its possibilities; the machine tool might be a lathe or a vertical axis miller (Fig. 3.3(a)). The NC vertical axis miller can be programmed to mill any complicated (2D) shape and controls the three axes of movement over the whole work table. Other examples are the furnace with programmed automatic control

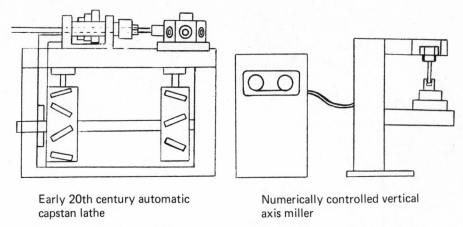

Early 20th century automatic   Numerically controlled vertical
capstan lathe   axis miller

Fig. 3.3(a) – Automatically controlled machine tools.

of temperature, the dish washer and the clothes washer (Fig. 3.3(b)) – these
may be batch operation or continuous.

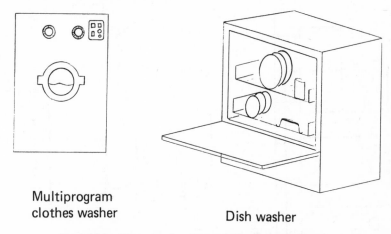

Multiprogram
clothes washer   Dish washer

Fig. 3.3(b) – The mono function variable program automat.

Level 4 is the man extender (Fig. 3.4). These are the machines which are
powered and carry out complex programmes for which they are instructed by a
man; but they only do these when a man is connected to them in 'real time,' at
least as far as telling them to do that particular job. The three examples in use
at present are: (1) the digital computer, which is capable of doing very quickly
any operation of very great complexity that can be completely represented by
numbers; (2) the telechir, which enables the skill of a human operator to be
exercised remotely from his body; (3) the powered artificial limbs controlled
by the action of other muscles of the handicapped person.

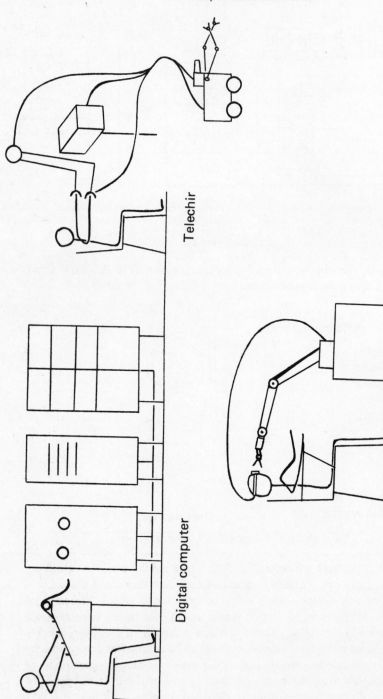

Telechir

Digital computer

Head-controlled artificial arm

Fig. 3.4 — Man extenders.

Level 5 is the robot (Fig. 3.5) — systems which are powered and programmed and will continue by themselves to repeat any one of a series of complex programmes given by a human master to produce a succession of identical products or carry out a series of similar operations as in storekeeping. It must have a memory for the series of operations it has been instructed to do, a method of instructing this memory by the human, and a mechanical manipulating device. On second generation robots it would also be possible for the human to instruct it to vary its actions in accordance with variations which it observes in the surrounding situation.

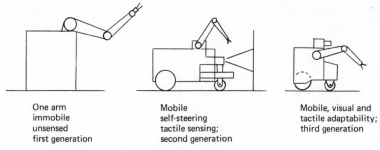

One arm
immobile
unsensed
first generation

Mobile
self-steering
tactile sensing;
second generation

Mobile, visual and
tactile adaptability;
third generation

Fig. 3.5 — Robots.

3.2 MAN/MACHINE CONTROL SYSTEMS

Broadly speaking all control systems can be divided into two groups, (1) simple switches (Fig. 3.6) which are on or off and (2) the analogue multi-step or continuously variable control systems (e.g. a rheostat).

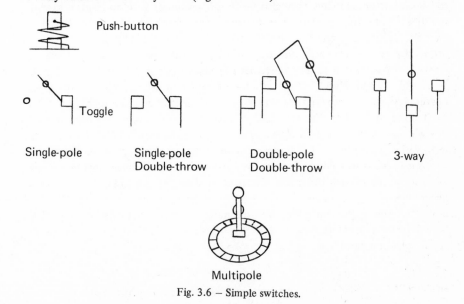

Push-button

Toggle

Single-pole

Single-pole
Double-throw

Double-pole
Double-throw

3-way

Multipole

Fig. 3.6 — Simple switches.

Switches may be arranged in complex keyboards, as in the typewriter or calculator, in which case they are push buttons each worked by one finger and the layout is very important. For example, in producing a light-spot operated typewriter for severely handicapped (tetraplegic) patients (*Aids for the Severely Handicapped,* edited by K. Copeland, Sector Publishing Ltd) the authors found it very much easier for the patient if the letters and numbers and signs were arranged in a series of concentric squares with the most commonly used in the centre. The mathematical centre at Amsterdam studied the frequency of occurrence of letter pairs in the Dutch language to minimise the distance travelled by the light-spot during typing. They put the three most frequent characters (e, space, n) in the centre and the next six (a, t, r, d, o and i) around them in such an order as to minimise travel distance for the most frequent pairs.

It is unfortunate that the standard typewriter keyboard, designed by C. L. Sholes[†] in 1873, is still used in spite of the fact that the layout is based on considerations of alphabetical order and of keeping type bars for common consecutive pairs of letter well apart so that they do not clash. Considerations of relative finger skill for touch typing using all fingers would have avoided placing O and P keys where the outer fingers of the right hand have to work so frequently while J and K are in the most easily used positions. Alternate hand-keying for the most common letter successions is also desirable in the optimum keyboard. In the new word processing systems automatic carriage return saves considerable time. Other keyboards have been developed in which the two hands operate two banks of keys arranged for hands sloping inwards from the elbows at the sides of the body and which make more use of the thumbs.

Push-button switches may also be foot-operated as with an electric sewing machine or the foot control of a magnetic tape recorder for typing. Switches may also be of the toggle type and in this case may have three positions with the off in the centre, forwards one way and backwards the other way or even a joystick on/off switch to give movement in eight directions round the compass as in the Queen Mary College system of powered wheelchair control. This operated two d.c. reversing motors, one on each side and the eight positions correspond to FF, FO, FR, OR, RR, RO, RF, OF respectively. In a later version this was improved to an analogue system in which the speeds of the two motors were varied according to the position of the joystick at any point on the circle. The speed was also controlled in proportion to the distance moved by the joystick in the chosen direction. When applied to the control of a robot or a telechir each switch would, in general, correspond to switching on the full power to one movement giving approximately constant velocity as long as it is held down, apart from the effects of varying inertia or loads and the acceleration and retardation periods.

[†]QWERTYUIOP – dinosaur in a computer age, I. Litterick, *New Scientist,* 8 January 1981.

Analogue devices control some variable continuously over wide range from zero to maximum as they are moved over a certain distance; the accelerator pedal of the motor is the best-known example. In the case of the spark ignition engine the accelerator pedal controls the opening of a butterfly valve in the air line while in the diesel car it controls the amount of fuel injected. In the case of robots and telechirs the movement may control displacement, velocity, force or power; frequently the control is via a sliding potentiometer. For controlling a single variable they may be knobs, cranks, wheels or levers (Fig. 3.7). A simple

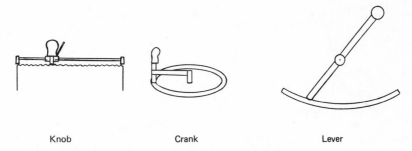

Knob   Crank   Lever

Fig. 3.7 — Single variable analogue controls.

joystick (Fig. 3.8) is frequency used to control two variables as it moves over the surface of a hemisphere or smaller fraction of a sphere. A third variable may be controlled by making the joystick telescopic, a fourth by making it sensitive to rotation about its own axis and a fifth by making some kind of pistol grip so that squeezing the palm controls it. Finally we have two systems whereby one hand of the operator can control all seven degrees of freedom needed for a telechiric hand, three displacements, three rotations and grasping. In one of

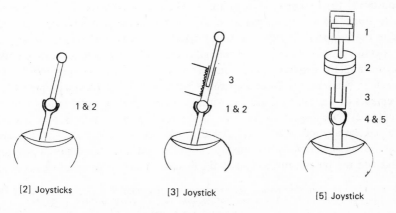

[2] Joysticks   [3] Joystick   [5] Joystick

Fig. 3.8 — Joysticks.

these (Fig. 3.9(a)) the human operator's hand grasps a pistol grip device which is supported by a master arm with six degrees of freedom which has exactly similar movements to the arm of the slave machine so that the seven movements of the man's hand are converted into movements of the master device and the movements of the master device are exactly copied by those of the slave device. In the other system the human puts his hand inside a 'glove' attached to an arm forming an exoskeleton to his arm so that his movements are directly monitored by this arm which is copied identically or on a different scale by the remote slave arm.

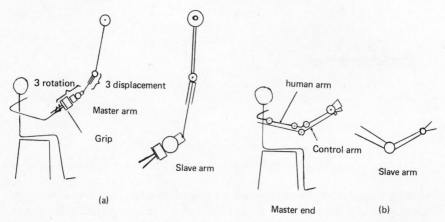

Fig. 3.9 — Telechir controls.

Another form of control worked out for telechiric devices is the 'resolved motion rate control'[†]. In this system the operator grasps a six degree of freedom, isometric hand control, and the force he exerts in the direction of each of the three displacements and rotations produce rates of movement in the slave proportional to the corresponding force. The actual movement of the control is very small. He also has a trigger worked with the forefinger for opening and closing the grippers. In this case a computer processes the commands so that the operator is not concerned with the interaction of the various slave moments.

Figure 3.10 illustrates the control system of a non-sensing robot (the Unimate)[‡] whereby the human operator instructs it to perform a long series of movements of the arm and gripper which it then repeats exactly as many times as required. Each degree of freedom of the robot arm is controlled by a servo valve and has a position encoder (digital or analogue). During the 'teach phase' the valves are operated by the human controller and when the required position is obtained the record button causes the memory to store the exact values of all

[†]*Information and control issues of adaptable and programmable assembly systems*, J. L. Nevins and D. E. Whitney, *Mechanism and machine theory*, 1977, Vol. 12, p. 27.
[‡]Ibid. p. 37.

the position encoders. During the repeat phase the hydraulic servo valves operate until the comparator determines that all the positions agree with those recorded in the memory.

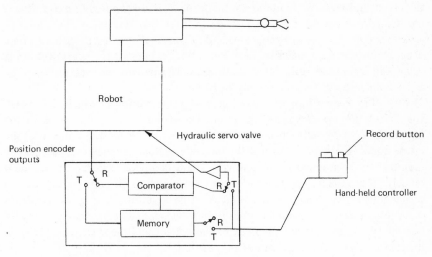

T = Teach position
R = Repeat position

Fig. 3.10 — Control for non-sensing robot.

A great deal of work has been done to enable paralysed or otherwise handi-capped humans to lead a more normal life by enabling them to use special kinds of switches and control systems to operate machines which do things which they would otherwise have done with their hands or their voice. The Possum device enables those whose only reliably controlled movement is blowing and sucking with their mouths to type and even work a coordinated table machine tool (vertical axis multi-station turret drilling head). It has been found that they can do 10 suction/pressure cycles per second and they can pick out the letter on a typewriter control by arranging the letters in a 6 × 8 grid with suck giving movement along one axis and blowing at the right angles to it (*Aids for the Handicapped*, p. 25 'Control systems — Concept and Development (POSSUM)', Reg Maling).

In another system, called the Pilot (*Aids for the Handicapped*, D. W. Collins, p. 31), a powerful light source on the console is connected by flexible glass fibres to a focusing device fixed to the patient's head so that he can point the light beam at an array of switches on the console. This array has two forms (1) an alphanumeric in four rows of eleven switches, and (2) an environmental control with six pairs of on—off switches. The light falling on a photocell switches on a thyristor which puts a d.c. pulse of 100 V to the solenoid which operates the appropriate typewriter key on a 'golfball' typewriter. Both systems can also be operated by a keyboard by multiple sclerosis patients with small but coordinated movements of the fingers of each hand.

The other light-spot operated typewriter ('LOT', M. Svede, H. G. Stassen and Avan Lunteren; W. J. Luitse, Netherlands *Aids for the Handicapped*, p. 42) has the light source fixed to a pair of spectacles so that movement of the head to look exactly at the control panel area and holding it there for a pre-selected time display operates the typewriter key. For different patients this preselected time delay can be varied from 50 to 500 msec. The patient receives feedback by hearing the letter typed and looking at the next typed.

Another system has been developed at Southampton in which the human voice (*Aids for the Handicapped*, p. 54, Morse code and voice control for the disabled', A. F. Newell) has been used to operate a typewriter by means of the morse code. Humans can control the duration of sounds much more accurately than the word sounds so a comparatively simple computer decoder can decode it reliably. The patient uses long and short sounds such as the classical 'dit—dah', a silence longer than 'dah' denotes a break between letters. A long gap of silence indicates the end of a word and the machine decodes and acts. A simple lanyard microphone is hung round the neck of the patient and the system can be used for controlling other things besides a typewriter.

In the case of people with cerebral palsy with strong compulsive movements of head and limbs it has been found possible in Finland to control a typewriter by means of a dental plate switch as there was good muscular control of the tongue and lips. This switch could be used for the morse code (*Aids for the Handicapped*, p. 59, 'Typewriter control by dental palate key', I. Saarnio). Another system that has been used is a capacitance probe to pick up body (e.g. head, shoulder, toe or hip) movements near a transducer plate (*Aids for the Handicapped*, p. 63). The probe switch operates a series of lights numbered 1—10 which select and indicate the condition of the 'logic state'. When the signal is stopped at a number that number operates.

Finally we have the electromyographic potentials produced by the brain control of a muscle. This may be picked up as a very small spike-shaped voltage pulse, for example by the standard electrodes used in the electrocardiogram. A. Torok (*Aids for the Handicapped*, p. 77, A typewriter operated by electromyographic potentials) used pairs of EMG electrodes on an arm muscle amplified up to × 1500, bandpass filtered (4—400 Hz) and subjected to a variable threshold to discard noise and produce logic pulses. These are counted up to 64 and operate an 8 × 8 array of indicator lights each of which is connected to a typewriter key or mains power switch. A second EMG signal from another muscle provides the trigger pulse to carry out the operation.

Morecki *et al.* ('Manipulators for supporting and substituting lost functions of human extremities', *First Symposium on Theory and Practice of Robots and Manipulators*, Vol. 1, p. 213) used the EMG signal from a single muscle to operate a six degree of freedom orthotic manipulator. They found it easier for the patient to operate the different movements sequentially than to use several muscles with separate EMG electrodes. The single muscle could give two levels of signals; the high-level signal selected and changed the type of movement while the low-level signal performed the movement. The amplifier and logic system were operated by 12 V d.c. and these operated the limb movements by electro-pneumatic valves controlling gas from a cylinder of liquid CO_2.

For many years studies have been carried out in the attempt to allow men to speak orders into a machine. In 1976 Reddy at Carnegie-Mellon (*Machines who Think,* Pamela McCorduck, W. H. Freeman, 1979, p. 266) produced a system which would accept a vocabulary of 1000 words with a highly artificial syntax for a special contact system continuously in the general American dialect. It has to infer correctly what the user wanted rather than determine exactly what had been said. G. V. Senin of Moscow ('National language for interaction with a database', *Machine Intelligence,* Ellis Horwood, 1979, p. 445) describes the 'dialogue information logical system' which has had its first tests in Russian and English. Like the US system they accept a restricted lexicon and context.

3.3 INFORMATION FEEDBACK FROM MACHINE TO MAN

When a man is working in conjunction with a complex machine he receives information from it by means of his eyes, his ears and by touch or proprioception; smell warns him of hazard. There are also sophisticated devices using electrodes on the skin or touch to the fingertips used for blind reading devices and for providing the feedback that a man might normally receive through his hand. Braille requires extremely delicate feedback of sensation of arrays of small dots through the fingertips.

By far the most important is the display which a normally sighted person receives through the eyes. This may take the method of reading pointers on dials as the speedometer of a car or the ammeter on a battery charger. Much work has been done on the arrangement of dials, for example anyone who has read an integrating watt hour domestic meter will know how difficult it is to read a series of dials when the alternate pointers move clockwise and anti-clockwise. It is very important also that the arrangement of a series of dials on an instrument panel, as in an aircraft, or the control room of a power station, is such that the most important ones are in the easiest place to see, but that if others further away could denote an occasional important error in the system then they should be combined with something to draw the attention of the eye to them, such as a warning buzzer. For very accurate scientific measurement the parallax (the error due to looking obliquely at a pointer located a significant distance in front of the scale) can be avoided by using a strip of mirror behind and only taking the reading when the head is in such a position that the reflection of the pointer is hidden by the pointer. Preparation of scales must be a good compromise between clear visibility on the one hand and inaccurately thick lines on the other hand. Scales for check-reading (e.g. car radiator temperature) are best marked with coloured bands, e.g. for low, safe, high, while graduated scales should be divided in decimal scales with longest graduations on the tens, medium on fives and shortest on units (*Man Machine Engineering,* A. Chapanis, Wadsworth, 1965).

The advent of solid state devices has made it possible to give a digital readout of many quantities previously only observed by dials, the best known being time and weight. These systems can obviously be very useful for auxiliary information in the case of telechirs such as the ambient temperature or the water pressure. By far the most important visual feedback system is the actual picture. This may be a two-dimensional or stereoscopic TV picture of the scene or it may be a reconstructed image or model drawn from visual data or from sonar or radar sensors and sensors of the position of the joints of the slave arm. In one system a TV screen has various areas giving information about what the human controller has instructed and what the system has carried out. While it is possible to have stereoscopic TV this has not so far been found to give a great increase in the realism of the man doing the work. A more effective system is enabling the TV cameras to move in response to the movement of the human master's head so that he can see the task from different positions, for example to look behind an obstacle. It may ultimately be possible to develop *remote moving laser holography* so that the man will have reconstructed in front of his head a complete three-dimensional image of the slave/hand and the task and he can move his head about to see from different angles while the slave/hand follows his own hand movements.

Audio feedback can be used in many ways in connection with telechirs, thus in mining, variations in sound of the cutters can indicate to the skilled operator when it moves from coal to rock, sounds can also give warning of machine defects as every car driver knows. It is also possible to put on special warning signals, for example when a slave-telechir reaches the end of its movement.

The human nose is incredibly sensitive in detecting for example the first indications of a fire. Sensitive ionisation detectors can give the same kind of indication but the sense of smell and taste are not so far used in either robots or telechirs.

Proprioception whereby the human receives feedback proportional to the force being exerted by a grip of muscles in the arms, legs or back and the position of the joint is of fundamental importance in telechirs, while tactile senses in the finger or hand can be of great value, both in telechirs and in robotics. There is no doubt that a skilled craftsman will not be able to exert his skill at the other end of a wire unless he can have at least some feeling for the force and touch position of the slave-hand. Since, however, it is very expensive to make each of these feedback signals, the basic problem is to know what is the minimum; thus many telechirs have been made with no such feedback but in this case it is essential to have a tactile sensor connected to a robot minicomputer control circuit on the slave so that the grasping operation of the hand in relation to the object is not carried out in such a way as to damage the object.

In the case of an artificial arm for an amputee (*Mechanisms and Machine Theory*, Vol. 12, p. 510) a sensory feedback system has been developed in which the amputee is given information both about the pressure on the prosthesis finger tips and the elbow angle of the prosthesis. The finger-tip pressure for prehension is monitored by a conductive rubber transducer which gives a frequency modulated signal to a pair of electrodes, modulated in the range 10 to 100 Hz. The elbow position is observed by a potentiometer and the signal is detected by passing the pulsing signal from one electrode to the next, for a movement of $20°$. There are eight electrodes arranged along a line or in a circle attached to the skin in the socket/stump interface.

CHAPTER 4

Sensors for robots, telechirs and aids for the handicapped

4.1 THE NEED FOR SENSORY FEEDBACK IN MACHINE SYSTEMS

We have seen in Chapter 2 the very complex sensory feedback that the human has trained to enable him to exercise his skills of movement and manipulation in the world. This involves the senses of sight (with all the sophistication of processing the huge amount of visual information) hearing, taste and smell, the proprioceptive sense from all the main muscles, touch sense from the skin (varying in sensitivity from the finger tip, which can be trained to read Braille, to large areas of the back with very few sensors) and the vestibular system which tells us about gravity and acceleration forces (e.g. rotation, which makes us dizzy). Clearly we shall never be able to build robots with more than a minute fraction of the human ability to receive and process information but in this chapter an account will be given of the experimental work which has been done to recieve and process information mechanically.

We start with vision. Here the problem of telechir vision is very much simpler than that of robot vision because in telechirics we have the use of the human eye and vision proeessing system. The attempt to convey to a blind person some of the information they are missing by using other senses or by direct electrical signals to the brain is also a very important field of work (section 4.5). The other main field of work is touch forces and proximity sensors (section 4.6).

4.2 VISION FOR TELECHIRS

The problem of providing effective vision feedback to the human master operating a remote telechir is much simpler than that of vision for a robot because if an adequate picture of the remote scene is presented to the eyes of the human master he can use the fully trained skill of his brain to interpret the scene. Some of the things he does are (1) to estimate the size of an object from the angle subtended and an estimate of the distance, (2) to coordinate his view of an object with his previous knowledge so as to be able to decide what to do with it, and (3) to control the movements of his hand to bring it into right relation to the object. If (4) a man tilts his head sideways he interprets the displaced image of what is seen as if it were still upright by using the information from his vestibular system to know the tilt of his head. The mechanism of these computations is shown diagrammatically in Fig. 4.1 In the third case he will generally use tactile information for the ultimate fine positioning. He will also make an estimate of the three-dimensional character of the object by moving his head round it, observing the shadows and if necessary turning it over or touching it. It has been shown that colour does not increase very greatly the information useful for a task but the direction of illumination to cast shadows and the vision from different directions can be very important. However, there are cases such as geological rock inspection or the use of colour-coded cables where colour would be important.

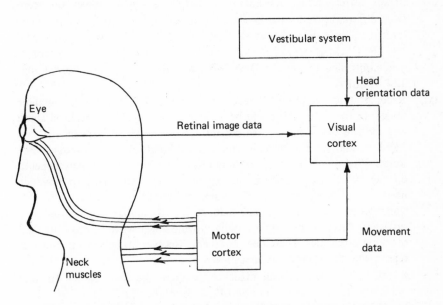

Fig. 4.1 — Simplified diagram of system whereby visual cortex corrects image to a fixed space: voluntary movement of eye or head.

There are basically six possibilities for remote vision for telechirs.

(i)   A simple black and white single TV screen fed by a fixed TV camera or a radar or screening system and viewed with both eyes. Alternatively the screen may show a model reconstructed by computer of the position of the telechir arm in relation to the viewing camera.

(ii) A single TV colour screen receiving signals from a colour TV camera.

(iii) Black and white stereo vision fed by a pair of cameras whose distance apart may be equal to or considerably greater than the distance apart of the eyes of the viewer. In this case a device such as spectacles with two colour filters or two filters passing light polarised at right angles is necessary to give the stereoscopic vision from a single screen to the two eyes of the master.

(iv) The use of two TV cameras looking at the work from quite different directions, for example with their axes at right angles to each other. These are fed back to two TV screens side by side and the master has to learn to interpret the information. The information is considerably more effective than a single screen, for example for putting a peg into a hole where it has to be located and orientated in two dimensions.

(v) A system in which the master can move his head about and the TV camera makes corresponding movements so that he is able to survey the scene from different positions as he would on a real task. This system is normally limited to the movements of his head which he can make sitting on a fixed seat (R. C. Goertz *et al.,* the ANL Mark TV2 — an experimental 5-Motion head-controlled TV system, *Proc. 14th Conf. Remote Systems Technology,* 124, 1966).

(vi) Reconstruction of the work seen by a three-dimensional laser holographic image. In this case, if it could be achieved, he could move his head about and see the situation in three dimensions as though he were on the spot and he could see the telechiric mechanical hands moving as though they were his hands which are controlling them. At present holography depends on the ability to produce a photographic plate with black spots which block one of the two split light beams and we do not yet have a way of transmitting such a holographic plate so that it can be reconstructed with sufficient detail to provide a rapidly moving image with the requisite light blocking properties. There does not seem, however, to be any fundamental reason why this should not be achieved.

A detailed study has been made to compare the first four of these six possibilities by L. A. Freedman *et al.* (*Mechanism and Machine Theory,* 12 (5), 1977, p. 425: 'TV requirements for manipulation in space'). They did a statistical evaluation of various operators carrying out simple tasks with a manipulator having four degrees of freedom in its mechanism, operated by four toggle switch controls. The task was in another room and could only be seen by means of the TV viewing system. The black and white TV camera was an industiral Grade 1 incorporating a silicon sensor while the monitor was a high quality 17-inch monochrome unit. The single picture colour system used a broadcasting camera and process system and a 17-inch good quality screen. The stereoscopic system used two black and white cameras with parallel axis separated by 6.2 inches. The two camera system provided a front view and a side view at 75° or 90° to the first camera. The tasks were relevant to various space activities.

They concluded that there was no strong overall effect in going from black and white to colour, that the two-view camera showed a significant improvement over the monoscopic system in the performance of the task, especially accurate positioning. The stereo did not give a significant advantage although it was better for some tasks. Thus the overall conclusion was that if the two-view system could be used it was the best, but if the extra complication was too great the simple black and white should be used.

A stereo system is used with a telechir serving a very large cyclotron (600 MeV protons) in Switzerland ('Minimac: the remote controlled manipulator with stereo television viewing at the SIN accelerator facility' *Proc. 26th Conference on Remote Systems Technology,* 1978, p. 62). The system reproduces two separate images onto the lens and TV camera by mirror prisms and polarised filters from two input lenses 12 cm apart so that 40 per cent of the screen surface is available for each picture. The convergence angle is remotely adjustable to focus on distances from 0.6 m to ∞. Repeated changes of viewing distance are found to cause operator fatigue but stereoscopic viewing of telechir manipulators without force feedback is regarded as essential to avoid damage to neighbouring components.

Another TV stereo viewing system is described in the same conference (p. 299) by H. W. Berry and O. F. Rice ('A three-dimensional television system for use in remote servicing'). One method of judging distance with monocular vision is to use the information from the eyes' focusing mechanism and this is not available with TV because the image is on one plane. In the United States TV during 1/60 sec 262.5 horizontal lines can scan the whole screen (field 1) and during the next 1/60 sec another 262.5 lines interleaved with the first ones also scan the whole screen (field 2). Signals from two TV cameras set a certain distance apart (corresponding to the spacing of the two eyes and set at an angle so that their lines of sight cross over at a certain distance are combined into a composite signal which is transmitted to the video monitor. Each camera provides one of the interleaved fields. The user wears a pair of spectacles which have have electronically operated shutters so that one eye sees each field. Any number of people can watch the same screen stereoscopically and they can move about while watching. The electronic shutters consist of two polarisers at right angles with a polarisation rotator between, which can be controlled by pulsed electric field.

An alternative stereo system ('Stereo television viewing for remote handling in hostile environments', B. Amos and M. Wang, *Proc. 26th Conf. on Remote Systems Technology,* 1978, p. 358) uses two small cathode ray tubes mounted on the sides of the head which are seen by the eyes via lenses and mirrors, and half-silvered mirrors in front of his eyes so that he can also see instruments or a fixed TV screen. The stereo TV camera is mounted on the 'shoulders' of the telechir, as is a separate colour TV camera. This system also provides the operator with the ability to pan and tilt the stereo TV camera by moving his head. When he finds what he is looking for he can press a button to stop the camera and then he can return his head to the comfortable straight position while the camera looks in the required direction. It is also possible to decouple the stereo TV camera from his head and manually rotate the camera ± 170° rotation ± 90° elevation.

A study of telechiric vision in connection with undersea telechirs is described by Charles and Vertut ('Cable control deep submergence system', *Mechanism and Machine Theory*, **12** (5), p. 481). They are working on a system in which the movements of the head of the operator will move a mechanical head on the slave carrying a miniature TV camera and two microphones. The platform has motions which relate closely to those of the human shoulders and neck and dynamics similar to those of the neck. The operator places the head in a helmet which has compensating springs to support its weight and that of the sensory stimulators, namely the earphones and the binocular TV screens. Signals of the man's head movements are taken from the supporting arms of this helmet system. Data have not yet been published on the improvements in skill obtained by this system but there seems little doubt that there will be a considerable improvement over the ordinary fixed head system and it will be much easier to use than the pair of mirrors at right angles because the information is presented to the user in precisely the form in which he receives it when he does the job in his own workshop.

A. L. Foote and R. J. Dompe ('Application for head-aimed television systems', *Proc. 26th Conf. on Remote Systems Technology*, 1968, p. 83) suggest that the reason head-aimed TV systems (HAT) enable operators to work efficiently with rapidly changing fields of view is that these systems make use of the operator's neuromuscular trained skill. By this his visual cortex is able to correct the image to locate it at the same position in space as he moves his eyes or his head and shoulders because it receives the necessary information from the motor cortex as to the position of the muscles controlling these movements (see Fig. 4.1). Operators who use fixed monitors are much more susceptible to discomfort or nausea.

The authors recommend the use of video displays giving a high resolution at the centre of the image where the eye has great acuity over an area of ± 5° from the fovea while there is a low resolution peripheral view to ± 30° to give the operator the use of 'the corner of his eye' to catch movements that may require the transfer of his attention. This is done by superimposing the images from two separate lenses.

It is also found to be important to have camera and eye positions in the same relation to their rotational axes if the sensation of movement at different distances is to be conveyed accurately.

Other requirement for a remote driving and manipulative telechir are quick response camera servos for azimuth and elevation, 875, or more, line resolution and remote foveal zoom.

The authors concluded that in spite of the extra cost of the HAT system it should be used whenever real time visual feedback is needed for telechiric systems where locating, orienting or manoeuvering in a remote environment is required. Fixed TV cameras are only adequate for manipulation if no real time orienting of views or manoeuvering of object or telechir are required.

It would appear that if telechiric manipulation is to give the trained crafts-man reasonable speed and tool handling adaptability for unexpected repairs then it will be essential to develop head-aimed television with body movement also to a standardised economical form.

4.3 FIBRE OPTICS FOR REMOTE VISION[†]

If a very fine fibre of glass or plastic receives light at one end it will be transmitted (with some attenuation) to the other end by total internal reflection at the surface which is usually a protective coating of much lower refractive index. When an incoherent bundle is used it transmits a single signal of light frequency (λ = 0.4–0.8 μm; frequency 0.13–0.26 $\times 10^{-15}$ Hz) which can be used over distances of several kilometres to transmit signals (preferably digital) of very great information content because of the very high frequency (see *Physics Bulletin*, Jan. 1975, p. 14).

A coherent bundle with the fibres grouped in the same pattern at each end can convey a picture of as many dots as there are fibres. Typically a bundle of 50 μm diameter fibres transmits 50 per cent of white light a distance of 1 m and about 40 per cent over 2 m. The maximum resolution of a tightly packed fibre bundle of fibre diameter d mm is $1/2d$ lines per millimetre.

Coherent bundles of overall diameter as small as 3 mm have been used as endoscopes to view inside human organs. An outer incoherent group of fibres carries light in and the object is focused with a lens system on to the end of the coherent bundle which can be viewed by the surgeon at the other end.

4.4 ROBOT VISION

The tasks which a robot may be asked to do in a factory, warehouse, mine or farm can be considered on a spectrum of complexity or sophistication.

(1)   The robot without sensors which repeats a complex series of manipulative movements exactly as it has been instructed regardless of whether the object to be handled is there or not and regardless of accidental obstructions. A mobile robot without sensors can only move on a preset path.

(2)   Checking the presence of the object to be manipulated which must have exact position and orientation. Similarly for a mobile robot observing the

---

[†]'Fibre optics in medicine and surgery', P. J. Brand, *Biomedical Engineering*, Dec. 1973, p. 508.

presence of an obstacle on the path which must be circumnavigated. Simple micro switches which switch to a different routine can be used in this case.

(3a)   *Sorting* objects of a small number of clearly different predetermined types or quality control of visible errors. In both cases the objects are silhouettes and always lie flat on the same face, but can have any orientation or they can be mechanically orientated.

(3b)   *Position fixing* with flat objects on a table or slowly moving belt. Again the object can have random orientation or in some cases can be mechanically orientated. The position of a hole into which a pin must be inserted is another example.

(4)   Three-dimensional objects of random orientation in a bin or resting in contact and overlapping on a table or belt — one object must be selected, picked up, and orientated and connected to a chassis or body. The human can screw it on.

(5)   *Terrain adaptation* for a mobile robot — climbs over obstacles, crosses crevasses, climbs ladders. This is as far as robots can be expected to go. All problems requiring the estimation of probabilities based on experience, judgement of best course with insufficient information or inventing a solution to a problem which cannot be solved by the obvious means (e.g. a rusted-in screw, or having to make a special tool, or encountering an unexpected type of obstruction) will never be done by robots or artificial intelligence. Many of them require the human emotional brain which, as has been postulated in Chapter 1, cannot be artificially constructed; others could be done but at an expense of construction costs that makes the telechir the only feasible solution.

The function of an optical sensor on a robot carrying out tasks in a fixed place in a factory can be divided into three groups.

(1)   The detection of the presence of a work piece or obstacle in a certain place which may range from a laminar object brought against a stop by a moving belt, to the detection of the presence of a three-dimensional object placed randomly in a heap in a bin.

(2)   The inspection of the detected object to see whether it is the piece being sought at this stage in the program.

(3)   The inspection of the object for deviations from the intended size and shape.

Another whole group of vision problems occurs in the case of moving robots. They may have to go round obstacles, climb over barriers or select their path by visual inspection.

There are basically two ways of providing robot vision in a form which can be used by the robot for modifying its actions. In the first of these only a few light sensitive sensor units are used and they are combined with hardware such as a mask and mechanical scanning systems such as rotating mirrors to give the necessary information. For these, in general, it is necessary to change the hardware every time the object sought is changed. In the other type a scanning device such as a TV camera or a whole array of light sensors up to 10,000 or more are used to provide complex information which has to be processed in a fairly large computer in a way slightly analogous to the way the human brain processes the information from the eye.

One example[†] of the matched hardware system focuses parallel light beams coming from a black object and a black mask with a hole of exactly the same size and shape as the object. The black object is on a bright white background and therefore on the focal plane a black dot appears at the place where the parallel light from the black object exactly coincides with the hole in the mask. At all other points on the focal plane the surface is white. In principle the position of this black dot gives the coordinates of the object in the plane, the contrast between it and the surroundings gives the degree of correspondence between it and the mask. It is also possible to have several holes in the mask of different shapes and the blackest dot will correspond to the one which fits the shape most closely. The principle is used to find the position of an object on a table by using a $45°$ angle swinging mirror between the object and the mask and the rest of the focusing system. The X coordinate is found in relation to the direction of swing of the mirror, the Y coordinate is found by splitting the photoreceptor into two elements and finding when the moving dot is balanced evenly between them while the Z coordinate (distance perpendicular to the plane on which the object is placed), is found by having sensitive elements in front of and behind the focal plane. To find the orientation of the object the reference mask can be rotated.

The need for a mask was avoided in the system developed by M. A. Brent at Queen Mary College (see Fig. 4.2, unpublished report) who confined his study to laminar white objects on a black background illuminated from above. He used a small array of light sensitive receivers connected by means of optical fibres to semiconductor light-activated switches. He centralised the image on the image normalising sensor which was a ring by means of gimballed mirror arrangements with two rotation angles and the size of the object was measured by varying the zoom lens until the periphery image just fell into the ring photodiode sensor. The image normalising sensor used two concentric ring sensors with a small difference of radius and the zoom lens was driven until the signal appeared on the inner ring but not at all on the outer ring. It was then possible to determine the orientation of a non-circular object because the inner ring was divided into four sections and these could be rotated to locate the point where the image just came onto the inner ring. However, this system was clearly limited in the speed at which the mechanical process could be carried out to obtain the required information.

[†]E. Mühlenfeld, 'An optical sensor for locating objects without marks or contact', *1st CISM—IFToMM Symposium On Theory and Practice of Robots and Manipulators.*

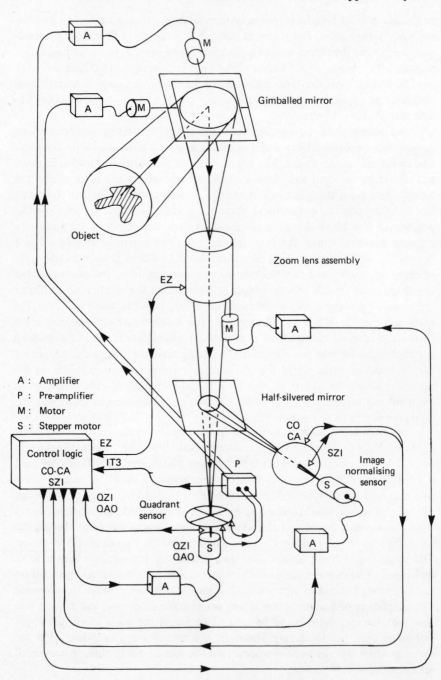

Fig. 4.2 – Schematic diagram of Brents' mask system.

An example of an industrial robot system using a TV camera is provided by the work of Heginbotham and Pugh ('The Nottingham SIRCH assembly robot', W. B. Heginbotham, D. W. Gatehouse, A. Pugh, P. W. Kitchen, and C. J. Page, *1st Conf. on Ind. Robot Technology,* 27 March 1973). In this robot (SIRCH) (Fig. 4.3) there is a four way turret located like the variable objective turret of a microscope; one of the positions of the turret puts an objective lens vertically above the table while the other three positions carry three different grippers. The turret can be moved about in three dimensions over the table and to simplify the calcualtions the TV camera can be rotated about this vertical axis mechanically to give the orientation of the object.

A 625-line TV camera is used with a 12K core store computer. The viewing area is 250 mm^2 and the component parts lying on the table are assumed to have a size range from 1 cm to 5 cm. The information is used from 128 points along each of the central 128 lines of the camera field. This system is used to scan the entire area for component parts and make preliminary identification of each component by indicating the area and perimeter of each component. When an object has been identified in this way the manipulator head of the machine is lowered to a point where the image of the object fills the whole television scan and the image is now compared with a reference area in the computer memory quantifying the various features of the object. The system used only a black and white image by placing the objects on a back-illuminated table so that the objects always appear black and no attempt is made to study the three-dimensional characteristics of the object, only its silhouette.

In another paper[†] Heginbotham describes the use of the same computer and program with a system of 128 photodiodes giving Y scan while the movement of the objects in one direction on the belt gives an X scan. This would be used in connection with a bowl feeder to detect the orientation and defects of flat objects. Two scanning systems can be used at right angles for three-dimensional objects. The components are characterised by the computer by four things.

(i)   The grid pattern, that is the boundaries of the object crossed by a number of parallel lines, e.g. four.
(ii)   The area and perimeter.
(iii)   The convex deficiency, that is the difference between the area of the actual object and the area of a string pulled tightly round it.
(iv)   External discontinuities of straight line segments, i.e. corners.

J. Bretschi, M. Konig and A. Schief ('Reduction of information optical sensors of industrial robots', *2nd CISM-IFToMM Symposium,* p. 285) have used a TV camera which gives a picture with 10^5 bits of information on one frame even

---

[†]A micro-processor controlled photodiode sensor for the detection of gross defects', A. Pugh, K. Whaddon, and W. B. Heginbotham, *Proc. 3rd Int. Conf. on Automated Inspection and Product Control,* April 1978.

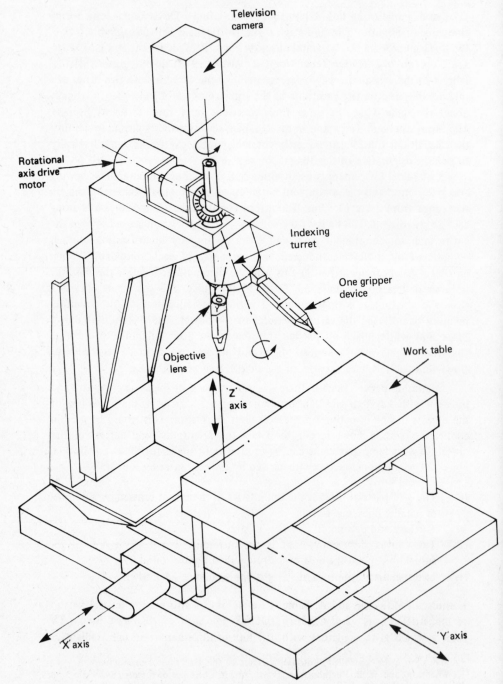

Fig. 4.3 – Heginbotham's SIRCH robot.

when each light spot only contains one bit, that is it is either black or white with no intermediate greys. They produce the contour picture by bringing a flat work piece on a belt, against a stop so that it has only a few stable orientations. There is either a fluorescent colour of the conveying belt with ultraviolet lighting or a transparent conveying belt and lighting from behind. A TV camera observes the intersections of a few lines (parallel to the stop) with the object. When setting it up for a new work piece a specimen is manually brought to the stop, in all stable positions, and the most useful lines of the frame are chosen by using a TV monitor to give intersections which are unambiguous between different positions of the work piece. Each intersection pair is stored in the form of a transition time in a Random Access Memory or a Read Only Memory. Five lines of a frame are sufficient for all tasks and there is a maximum of six intersection points for each line and eight bits for each intersection time so that the whole memory requires 240 bits and it can be stored on a single chip.

J. P. Foith, H. Geisselmann, W. Lubbert, H. Ringhauser ('A modular system for digital imaging sensors for industrial vision', *3rd Symposium CISM-IFToMM*, p. 399) describe a hardware and software system called MODSYS which is a modular system which can be used for the shape analysis of silhouettes of parts for several sensor tasks including sorting of work pieces, determination of position and optical inspection. They use a TV camera with flat black work pieces brought against a mechanical stop and scan by a number of lines with no grey scale (binary image processing). They use commercial TV cameras but are planning to adapt the system to solid state scanning plates. The following features are extracted from the object.

(i)   The area (by summing the size withing the silhouette).
(ii)   The centre of gravity.
(iii)   The maximum extension size in the vertical and horizontal directions.

These are all computed from the coordinates of the edges where the TV lines cross the black object. The programme can therefore be used for detecting variations in the area of a sequence of objects such as detecting a broken biscuit or the volume of axially symmetrical objects, such as a falling drop of glass. It can be used to determine the position, orientation and resting mode of a single object on a table. When the object has many different stable rest positions it can be positioned by processing the area and the two moment of inertia. The information is stored for a few positions and the observed values are used to interpolate between these positions.

One way of overcoming the problem of the variable distances of the work piece and TV camera eyes is to provide the camera with Fresnel optics in a system developed by P. R. L. ('Philips Research Laboratories aims at growth', *Electronics and Power*, Nov. 1980, p. 865). Two cameras are used, sighted roughly on the centre of the work table at right angles and these views are scanned electronically to enable the movable overhead suspended robot gripper to be located to an accuracy of 0.2 mm. The vision process is carried out by a minicomputer which also does strategic control of the manipulation but the detailed execution of the strategic commands to the manipulator is carried out by a microcomputer.

A. H. Bond ('Fast vision for a low cost computer controlled robot', *2nd Symposium CISM-IFToMM,* p. 273) uses a 100×100 photo array which can be scanned 10 times per second. The 10,000 levels of each scan are digitised to 8 bits fed into a minicomputer via a fast (4 MHz) A/D converter. A high level large time-shared computer is used to construct models and programmes while the on board minicomputer tracks the percept against the model.

The Stanford Research Institute automaton (Shakey)[†] uses a TV camera which has motorised focus, iris, pan and tilt controlled by on-board logic. The information is sent by radar or cable to a visual pre-data-processor and a large time-sharing computer. This can perform tasks in a real environment such as locating a special object of simple geometric shape and identifying it from other objects, going to it and pushing it to a specified position, if necessary going round obstacles on the way. The steps of processing the picture are as follows.

The total TV array is digitised into some 250,000 points which are essentially two-tone. The image is then differentiated or outline-enhanced to determine the boundaries of the lightness change areas. The white dots on this differentiated image representing the boundaries are then converted into a series of short line segments of three or four dots each segment. These line segments are grouped to form straight lines as far as possible and the image is converted into a line drawing of the outlines. From these line drawings the boundaries of the floor and the position of the objects on it and the shape and dimension of these objects can be computed.

A method for enhancing a faint TV image has been described by R. C. Gonzalez and B. A. Fittes ('Grey level transformations for interactive image enhancement', *Mechanism and Machine Theory,* **12**, 1977, p. 111). They start with an image digitised in an $N \times M$ picture element array (called PIXELS) and each PIXEL has a grey level in the range between 0 which denotes black and 1 which denotes white. We wish to make the contrast of the picture clearer by converting this grey level X to a transform Y where Y is also between 0 and 1 and the transformation function $(Y = T(X))$ is single valued and strictly monotonic. We call the probability density function in the original image $\rho_x(X)$ and in the transformed image $\rho_y(Y)$. A standard method known as histogram equalis-ation is defined as $Y = T(X) = \int_0^x \rho_x(S)\mathrm{d}s$. Clearly a very poor image is one in

[†]*Thinking Computer,* B. Raphael, Freeman, 1976. 'A mobile automaton', N. J. Nilsson, *Int. Jt. Conf. on Artifical Intelligence,* May, 1969.

which all values of X are in a histogram with a sharp narrow peak while one giving a much more uniform spread, will give clearer contrasts, especially if these widely varying values of the greyness are spread over the whole image. They develop a method which gives greater improvement than histogram equalisation because it can increase the contrast in that part of the image where it is most important.

Many people have worked on the development of computer systems which will compare with the human eye in being able to read print of different styles and different quality and a major project was going on at the NPL in Britain 15 years ago.[†] It was even hoped that it would be possible to read hand-written postal code block characters by a machine and sort them automatically. So far, however, this has not been successful and machine reading is confined to magnetic printed type as on cheques which has been made to look approximately like ordinary printed numbers but with greatly simplified and standard magnetic code.

The Wabot two-legged walking robot at Waseda University in Japan ('Information Power Machine with Senses and Limbs' *Theory and Practice of Robots and Manipulators,* 1st CISM-IFToMM Symposium Vol. I, p. 11, by I. Kato *et al.*) has two TV cameras in its trunk which can rotate to scan the space in front of it when searching for an object. The robot is connected by cable to a large computer which controls all its movements. When the scanning camera detects an object it emits a signal which stops the trunk rotating and notes the distance and angle. The camera lens can be automatically altered to improve the accuracy of measurements so that the measuring error of distance is 7 cm at 5 m and 4 cm at the closest possible distance of 0.6 m. The robot then sets off to walk to the object which it has detected. Sixty-four scanning lines are available for detection and the one selected gives the vertical height of the object, the trunk rotation gives horizontal direction and the use of the two cameras gives the distance. The data are processed as far as possible by hardware to enable a minicomputer to be used.

Work now proceeds in artificial intelligence laboratories on simpler pattern recognition compared to the hope a dozen years ago[‡] that we would one day develop a robot that could distinguish photos of a man and a woman. Industrial robots are not likely to have visual powers much beyond the ones at present in the development stage described above.

4.5 VISION AIDS FOR BLIND PEOPLE

Many devices have been developed whereby a blind person can point a torch or walking stick forward and receive an audible or tactile signal which will tell him how far away is the solid obstacle at which the stick is pointed. For example,

[†] 'The 2-ness of a 2', J. R. Parks and D. A. Bell, *New Scientist*, 20 June 1968, p. 624.
[‡] *Science Journal*, October 1968.

L. Kay (Lanchester College of Technology)[†] has been working on the use of ultrasonic mobility aids for the blind. A hand-held torch weighing about 0.26 kg is attached by a lead to a battery and hearing-aid ear-piece. Two distance ranges of 2.1 m and 6 m can be selected by push-button. The beam width is 15° and the mean transmission frequency is 60 kHz. The frequency sweep is such that when the beam is echoed by an object at the chosen range there is an audible echo note of 3 kHz. The pitch is proportional to the distance. He has also experimented with a binaural system in which the transmission frequency is made to vary in a sawtooth manner over a wide frequency band of up to 1 octave and two spaced transmitting and receiving transducers cover an area of 60°. When the obstacle is at an angle to one side or the other the echo notes to the two ears will differ in frequency as well as interrupting at slightly different times.

A tactual substitute for sight developed at the Smith Kettlewell Institute of Visual Science (*New Scientist,* 27 March 1969, p. 677) enables a blind person to lean his back against a dentist's chair which has a 20 × 20 matrix of 400 small vibrators. The blind person can manipulate at will a tripod-mounted TV camera. The signal from this camera is sampled digitally to operate a certain number of vibrators so as to give a crude tactual version of the image picked up by the camera. A 20 × 20 matrix is a poor substitute for the million or so points of the human retina but nevertheless it is a valuable instrument for blind people. Another system which has been studied experimentally in Britain (MRCs Neurological Prostheses Research Unit) and America (Institute of Bioengineering, Utah) has implanted an array of 64 electrodes in the visual cortex of the brain of a blind person. These implanted electrodes are simulated to give impulses of an appropriate pattern. The patient can see small white dots corresponding to the electrodes which are simulated, in the British case by means of radio waves to a receiver planted between the scalp and the skull; in the American case wires actually pass through the scalp. The receiver planted beneath the scalp could not be serviced, so this method was later abandoned.

An electronic 'human eye' is being developed by Dr W. H. Dobelle – director of artificial organs, Surgery Department, Columbia University (*IEEE Spectrum,* September 1980, p. 89; *Neurosurgery,* October 1979, p. 521). The plan is to have a micro TV camera in a glass eye attached to the eye muscles. The signals are processed in a microcomputer on spectacles and the processed signal operates 1024 electrodes on a Teflon wafer implanted on the visual cortex. The surface area of the visual cortex averages about 4 cm² and the phosphenes (light-sensing dots on the cortex) do not correspond in a regular pattern to the image so that the blind person's cortex has to be mapped to find how to connect them and a read-only map will have to be put into the computer.

A multi-pin carbon connector which can pass through the skin without fear of infection has now been developed.

[†] L. Kay, 'An ultrasonic sensing probe as a mobility aid for the blind', *Ultrasonics,* **2,** 1964, p. 53.

P. L. O'Donovan and L. F. Lind ('A fibre optic direct translation reading aid for the blind', American Foundation for the Blind: *Research Report,* 1977) have studied a system in which eight sensing fibres (each with light-transmitting fibres on each side) scan the print and produce a moving 'picture' on a 8 × 8 array which can be made tactile for blind people. The pair of light transmitting fibres for each sensing fibre are illuminated by a rotating shutter between their ends and a powerful lamp. The sensing fibres are divided into two interleaved groups of four, each group going to one detector and amplifier and the amplified signals are sorted out in relation to the shutter rotation to give a correct pattern on the display. Horizontal scanning is produced by physically moving the reading head across the page.

In the third CISM-IFToMM Symposium on Theory and Practice of Robots and Manipulators (1978) a preliminary discussion was given of a guide dog robot for the blind ('Guide dog robot, its basic plan and some experiments with MELDOG Mark I', S. Tachi *et al.,* Mechanical Engineering Laboratory, MITI, Tokyo). So far they have built a three-wheeled robot which propels itself and steers itself by means of a driven rotable front wheel. Ultimately this will be able to detect obstacles and inform the master or go round them and he will be able to maintain himself at a fixed distance of about 1 m behind the guide dog by means of ultrasonic communication from an oscillator on his belt to a receiver mounted above the back of the 'dog'.

4.6 TOUCH, FORCE AND PROXIMITY SENSORS

Much work has been done in the area of providing the equivalent of the human touch and muscular sensors of force (proprioception) to robots and telechir slaves. This work ranges from simple contact switches in the finger-tips of a hand (Fig. 4.4), such as I made 20 years ago, to close until it grasped a cylindrically symmetrical object of any size[†] to highly sophisticated multiple tactile systems. They can be divided into five groups:

(i)   The measurement of the principal vector set, that is the three perpendicular forces and three torques, for example on the wrist of a robot hand.

(ii) Tactile forces corresponding to the touch sensor in the human finger-tip. These may be of two kinds, a single touch and a matrix of contact points giving information about two- or even three-dimensional shape.

(iii) Measurement of inertia forces.

(iv) Proximity sensors: devices in which the presence of an object within a small distance of the sensor gives a signal.

[†]By using micro-switches backed by springs of different strength in different finger-tips it could grasp with different strengths.

Fig. 4.4 – Three-fingered hand with contact switches.

Two sophisticated systems have been developed for the measurement of the principal force vector. Flatau (Force sensing in robots and manipulators, *2nd CISM-IFToMM Symposium,* p. 294) describes a six-axis wrist sensor in which two concentric rings are used, one attached to the end effector or hand and the other to the supporting arm. They are cut away in a complex fashion so that six strain gauges attached to narrow bridges on them can be used to give the three forces and three torques.

J. L. Nevins (*Advanced Automation Systems and Manipulation Robots Course, Toulouse,* 20 September 1976, Vol. 2, p. 1) has described a wrist force sensor (Fig. 4.5) consisting of two rigid rings joined by three light strips each carrying an extensional strain gauge and a strain gauge shear bridge. From these six readings the three forces and three torques can be evaluated. The device is calibrated by applying the three forces and the three torques to it by means of weights and pulleys. It is to be used in conjunction with a compliance spring for fine assembly work such as inserting a pin into a hole which it fits with only a small clearance.

Umetami (*Tokyo Institute of Technology Résumé of work on Systems and Bioengineering,* 1980) describes Piezo-electric tactile sensor for a micro-manipulator for carrying out work under a microscope and for micro-surgery.

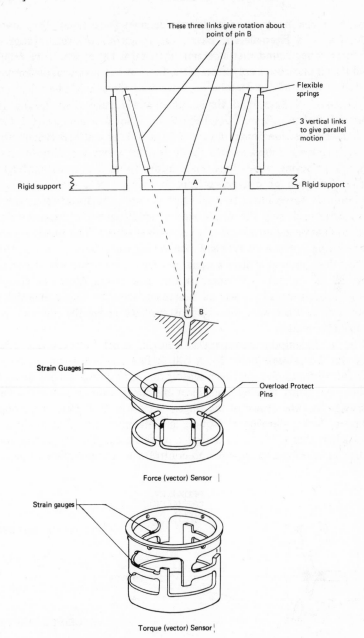

Fig. 4.5 — (a) Compliant wrist for inserting pins (P. C. Watson *Industrial Robots*, Vol. 1, p. 404). (b) SRI Strain Gauge Wrist Sensor (six-axis).

The sensitivity can be as good as 10 microgrames force (μg_f). They use the self-oscillation of a Piezo-electric beam, the frequency of which varies with contact force. They found that a ceramic material is better when one required rapid and firm motion but a polymer is better for slow but dexterous performance.

Heginbotham *et al.* ('A method for three-dimensional part identification by tactile transducer', N. Sato, W. B. Heginbotham and A. Pugh, *Proc. 7th Int. Sym. on Industrial Robots, Japan,* October 1977), ('Novel techniques for tactile sensing in three-dimensional environment', J. Page, A. Pugh and W. B. Heginbotham, *Radio and Electronic Engineer,* **46**, 1976) describe a very sophisticated tactile array capable of giving three-dimensional information. It does not suffer from perspective distortion or distance effects. It consists of a block carrying an array of iron rods, for example 16 x 16 which is lowered in discrete steps towards the table. After each step the whole area is rapidly surveyed electronically to see if any rod has been moved with contact with an object. This survey is carried out by energising one row of rods and then sensing each colum in turn so that if one rod has been moved it gives a signal and then the second row is energised and again all the columns are sensed in turn and so on. After one complete sensing the device lowers by a second increment. Thus the height at which any rod comes into contact with the object is available giving the required three-dimensional information.

Gill *et al.* ('Computer manipulator control, visual feedback and related problems', *1st Symposium,* paper 23, A. Gill, R. Paul and V. Scheinman) describe a robot hand with switch-type touch sensors on the two fingers based on rubber pads. When one finger touches the object the hand stops closing and a visual check is made that the centre of the fingers is close to the centre of the object. C. R. Flatau, 'Force Sensing in Robots and Manipulators' (*2nd Symposium,* p. 294) suggests that it would be desirable to determine directly inertial forces on a robot or telechir arm in order to control it to a precalculated trajectory

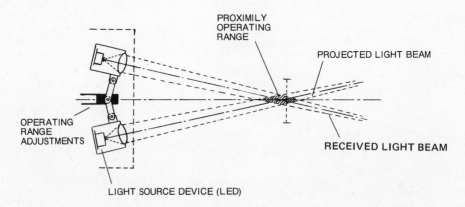

Fig. 4.6 – Proximity sensor.

rather than take derivatives of position or velocity measurements. He suggests the use of a pair of matched accelerometers on each link placed symmetrically around the centre of rotation of the link and connected additively so that all other forces are balanced out. A. R. Johnston ('Proximity sensor technology for manipulator end effectors', *Mechanism and Machine Theory,* 1977, p. 95) has developed a proximity sensor based on a light emitting diode illuminator and a compatible detector in which the sensitive volume can be adjusted by changing an external lens (Fig. 4.6). Detection is possible up to a distance of 1–2 m and the sensitive volume can be as small as 0.5 x 3 m or it can be made considerably larger. The presence of the object is detected by the diffuse back-reflection of the light emitted, the system being so arranged that light from the diode only reaches the detector when there is such back-reflection from near the focal point where the optic axes of the light and detector systems converge. The use of four such sensors, two on each jaw and a vice-like hand could be used with a control system to give three-dimensional control and some control of finger action.

CHAPTER 5

Mechanical arms and hands

5.1 MECHANICAL HANDS FOR ROBOTS, TELECHIRS AND THE HANDICAPPED

Theoretically, to grip a two-dimensional object requires three points of contact and a three-dimensional object requires seven points of contact. If the objects are appropriately shaped these grasping points could find a position where the three degrees of freedom of movement for the two-dimensional object or the six degrees of freedom of movement of the three-dimensional object were all eliminated by non-frictional supports acting normally, for example a three-pointed star held by three forces acting on the re-entrant corners or, in the case of a three-dimensional object, a tripod with three hemispherical feet resting in three V-shaped grooves meeting in a point in one plane with the seventh force pushing the tripod down on this plane. However, in other cases friction will be needed to prevent rotation, for example with a circular disc or a sphere. Thus, all mechanical hands, no matter how many points of contact they have, are to some extent dependent on friction.

Bianchi and Rovetta ('Grasping process for objects of irregular shape', *3rd CISM-IFToMM Symposium on Theory and Practice of Robots and Manipulators*, p. 67) have studied the two-dimensional case of a mechanical hand with three points of contact on irregular flat heavy objects in a vertical plane. Two of the fingers have friction (coefficient 0.2) and one is a frictionless horizontal surface pushed downwards by a spring. The two with friction have cylindrical contact points and can be brought by multiple linked fingers towards the object which is then pushed by them up against the sprung frictionless plate. The condtion of static stability is that the point of action of the resultant of the downward mass and the normal force on the horizontal upper plane contact point lies within the quadrilateral bounded by drawing the two friction angles on each side of the normal force from the cylindrical fingers. If it lies outside there will be rotation because the friction at both fingers will not be sufficient to prevent it.

M. Mori of the Institute of Industrial Science, Tokyo University (*New Scientist,* 4 April 1967, p. 282, *Four-finger exercise,* D. Fishlock) has analysed earlier studies on the role of the human hand on different jobs obtained by studying the distribution of callosities. Two hundred different crafts were analysed and it was found that nearly 60% of the tasks required by the hand needed only the fingers for their execution, while co-operation between finger and palm accounted for another 30%. The most important single action is the 'pinch' of the thumb and first finger. As a result of this study he designed mechanical hands having two or three fingers and a thumb, each having three joints, all under the control of a computer.[†] The human hand has four degrees of freedom for each of the four fingers, three curling and one sideways and has five for the thumb because each of the two lower joints has two degrees of freedom; this is what enables the thumb to oppose the fingers or work for them, for example it can curl round a handle the same way or in the opposite way to the other fingers. Thus the human hand has overall 21 degrees of freedom, not completely independently controlled but at least to some extent. Some people can bend the top joints of their fingers while holding the middle joint straight; certainly the forward bending of the middle joint and the basal joint are independent.

F. R. E. Crossley and F. G. Umholtz ('Design for a three-fingered hand', *Mechanism and Machine Theory,* 1977, **12**, p. 85) have also studied the manipulative movements of the human hand from the point of view of designing a hand for a telechir and a prosthetic device. They conclude that the two most important manipulations (apart from grasps) are to be able to pick up a tool (e.g. a screwdriver) and draw it into the nested grip against the palm, and to be able to hold a pistol-grip tool such as an electric drill and pull the trigger.

Some of the grasps of the human hand are holding a cylinder, a sphere, a luggage handle, a pill between two fingertips, a pen, a folded newspaper or a plate. More complicated manipulations such as threading a needle or folding a sheet will always be outside the possibilities of a multi-purpose robot and even probably a telechir.

Their primary objective was a telechiric hand for space manipulation viewed by a TV camera, so they wanted the hand as anthropomorphic as possible to use the operator's manual skill as though the hand were his own. The manipulations they wanted to do were those requiring the independent movement of the index finger (trigger and switch operation), the nested grasp in the palm of wrench or screwdriver followed by exertion of a torque and the use of cutters.

They concluded that they should design a hand (or, as they called it, 'end effector') with three digits, a thumb and two fingers. The third finger was separately motorised for trigger action. The hand is shown in Fig. 5.1. The

[†]M. Mori and T. Yamashita, 'Mechanical fingers as control organ', *JSME Conf. Proc. Session 3, Paper 4,* p. 106.

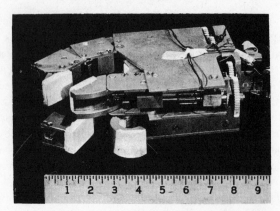

Fig. 5.1 – Three-fingered hand.

fingers were bent by working the three joints simultaneously with a single
flexible cable that drove right- and left-handed screws in each joint simultaneously.
The thumb and fingers could be operated separately or together, forwards or
backwards, by using preselection switches and 'run' switches. A thumbnail with
separate advance and retraction was also available. The thumbs and first finger
have a parallel jaw working surfaces operated by a set of parallelogram four bar
linkages in series. There is also a fixed cylindrical palm grip surface.

R. S. Mosher's gear-driven slave hand is shown in Fig. 5.1(a).

The various mechanical hands which have been constructed for robots,
telechirs and handicapped people may be divided into the following categories
of increasing complexity.

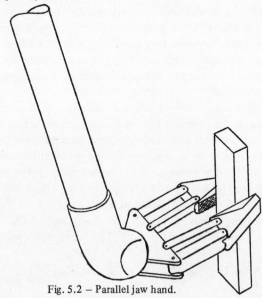

Fig. 5.2 – Parallel jaw hand.

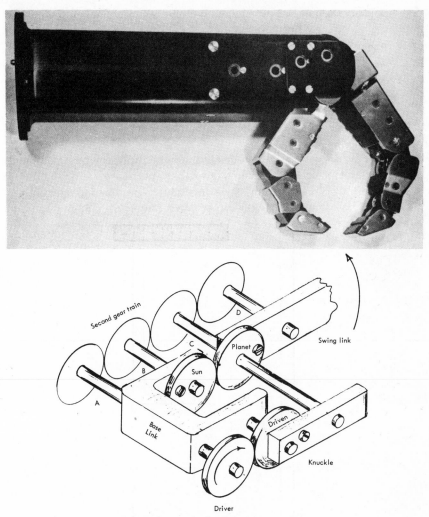

Fig. 5.1(a) – R. A. Mosher's two-fingered hand.

(i) The simple two-jaw vice-like hand or gripper. The jaws may have a parallel motion with link system to keep them parallel, Fig. 5.2, or they may have fairly long arms rotated about fixed points. The jaws always have to have high friction pads just as the jaws of a mechanical vice are heavily knurled to give a grip. If they are essentially flat jaws and have no ability to spring then they can grip an object with perfectly parallel flat faces in six contact points but a wedge-shaped object is only gripped in four and a conical object in two. This may be increased by making one of the faces spring to rotate about two independent axes, for example by supporting it with one fixed support and two springs. It is also possible to design a two-jawed hand to give six points of contact on a cylindrical object by making the jaws as in Fig. 5.3 ('Mechanised assembly system', W. B. Heginbotham, *Indian Machine Tool Design and Research Conference*. 1971). In this system, one jaw consists of a single V-shaped narrow plate and other of two V-shaped narrow plates. This type of system can, however, only grip cylindrical objects with the axis perpendicular to the jaw plates, or if they are above a certain size then in a perpendicular direction, with the six points of the Vs.

(ii)   Three-jawed systems. These systems are essentially for grasping laminar objects or objects generated by a single lamina moved at right angles to itself. That of Bianchi already described has one finger sprung while that of H. Hanafusa and H. Asada ('Adaptive control of a robot hand with

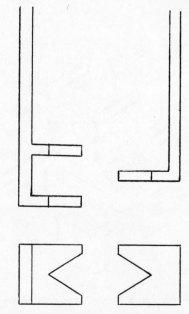

Fig. 5.3 – Heginbotham's two-jaw hand.

elastic fingers for mechanical assembly', *3rd CISM-IFToMM Symposium*, p. 45) has three sprung fingers independently driven by step motors; each finger has a twisted coil spring and the finger displacement can be measured by a potentiometer connected to the fingertip while the finger force can be measured by a potentiometer connected across the spring. This hand is intended for assembly movements and the springiness of the fingertips is primarily to give compliance when the object is not perfectly aligned with the hole into which it has to be inserted.

Figure 5.4 shows an experimental two-jawed hand or vice with two disc-shaped jaws each having eight shallow V grooves in the face and one sprung with three strong springs to rotate about a central ball by a limited angle. This can grasp conical, cylindrical or even large spherical objects as the unsprung disc is concave at the centre.

Fig. 5.4 — Two-jawed gripper for objects of various shapes.

Figure 5.5 shows an experimental hand with two jaws in the form of flat plates with diamond-shaped grooves and with one jaw sprung on four springs to grip irregular objects. It also has a third finger separately operated to push an object into the jaws.

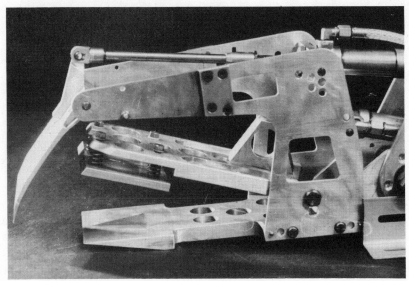

Fig. 5.5 — Two degrees of freedom hand.

Figure 5.6 shows a two-jawed hand requiring a single movement to pick up and grasp an irregular object lying on a table. One hand consists of a 60° V, making an angle of 15° with the lower surface which rests on the table. This contains grooves through which the seven sprung fingers of the other hand can slide as it rotates and scoops the object into the apex of the V.

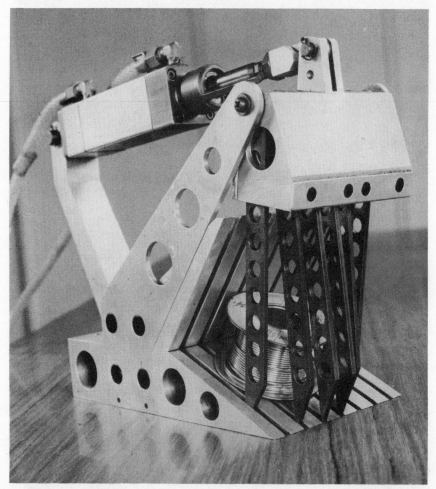

Fig. 5.6 – Two-jawed hand for picking and grasping objects of various shapes.

(iii)   A hand with three fingers and a palm, see Fig. 5.7. This was designed primarily for grasping objects like screwdrivers. The movement of the palm forward on a screw thread by a motor rotating the nut also brought the three fingers together with a concentric movement. The palm was in the form of a hollow cone and the fingertips were equipped with micro-switches so that the object could be grasped with a strong or a weak grip according to the strength of the spring in the fingertip.

Fig. 5.7 — Three-fingered hand.

(iv)   Human-like hands with an opposed thumb and two or more fingers. These have been developed both for cosmetic reasons, when making artificial hands for the handicapped, and also because of the astonishing adaptability of the human hand. They often have finger-joints that curl together. One such system (M. S. Konstantinov and Z. I. Zarkov, 'A kinematical algorithm and dynamical point mass simulation applied in robots and manipulators' *1st CISM-IFToMM Symposium,* Vol. 1, p. 69) has inflatable fingers, each one being a hollow half bellows of glass fibre reinforced urethane. Air pressure causes all the fingers to curl inwards. However, hands require several independently controlled drives if they are to do more than clasp cylindrical objects of different diameters. Those of Crossley and Mori have already been mentioned, another one is described by A. K. Bejczy ('Allocation of control between man and computer in remote manipulation', *2nd CIMS-IFToMM Symposium,* p. 417). This was originally designed in Belgrade as a prosthetic hand and has all the joints of a human hand except that the thumb cannot be aligned with the fingers. In this work the control was modified so that the operator or a computer programme could limit the grasping action to either the fist or the pinch mode, that is with all the fingers curled round the object, with the thumb opposed, or between the thumb and the first finger only.

(v)   The soft gripper. Y. Umetani (Professor of Systems in Bioengineering, Tokyo University of Technology)[†] has described a mechanism (see Fig. 5.8) which can softly and gently conform to objects of any shape gripping them in a single plane and hold them with uniform pressure.

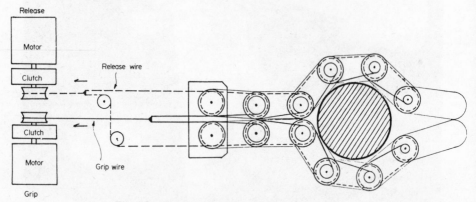

Fig. 5.8 – Total mechanism of the soft gripper.

Each link carries a pair of pulleys, running freely on the joint axis. The grip wire runs from each larger pulley to the next smaller one and the smaller ones get steadily smaller towards the last joint. When the grip wire is pulled it produces a closing torque about the nearest joint until that segment is stopped by an obstacle whereupon the tension is transmitted to the next joint. By making the inner pulley successively smaller the grip force on each segment is equalised. The jaws can conform to concave as well as convex surfaces and by making them at a slight angle up and down they can overlap right round an object as in Fig. 5.9.

5.2 THE DESIGN OF MECHANICAL ARMS

5.2.1 Kinematics

There are six possible mechanical joints for arms, and arms may be made up of various numbers and combinations of these joints. These joints are illustrated in

[†]Y. Umetami and S. Hirose, 'Biochemical Study of active cord mechanism with tactile sensor', *3rd Conference on Industrial Robot Technology, Nottingham 1976,* paper C1–1, 'The Development of soft gripper for the versatile robot hand', *Mech. and Machine Theory,* 1978, **13,** 351–359.

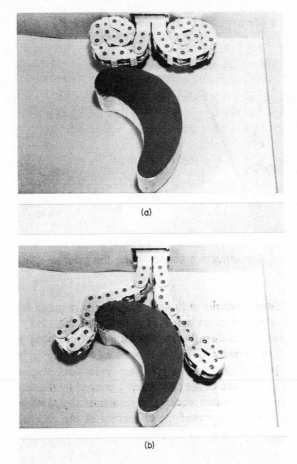

(a)

(b)

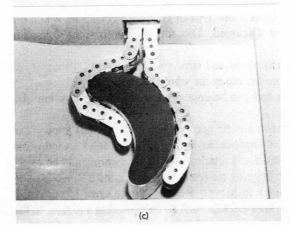

(c)

Fig. 5.9 – Soft gripper.

Fig. 5.10. The simplest is the simple hinge joint called by Roth[†] the *revolute* or R. The revolute joint can be in three different directions, in the most usual shown as R_1 the axis of revolution is at right-angles to both links; this is similar to the human joints. In the second type (R_2) the axis of revolution is along both joints so that the end rotates about the common axis. This is roughly equivalent to the human lower arm joint where the hand can be rotated about the axis of the lower arm. In the third type the axis of rotation again coincides with the upper arm but the lower arm rotates around it, at right-angles to it. Revolute joints can be driven (1) by rotary hydraulic or pneumatic actuators with limited angle of rotation, (2) by electric or hydraulic motors with total continuing

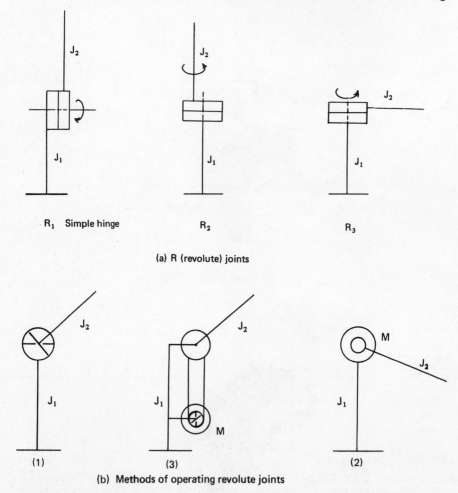

R_1 Simple hinge R_2 R_3

(a) R (revolute) joints

(b) Methods of operating revolute joints

[†]B. Roth, 'Performance evaluation of manipulators from kinematic viewpoint', *Cours de Robotique*, Vol. 1, p. 235.

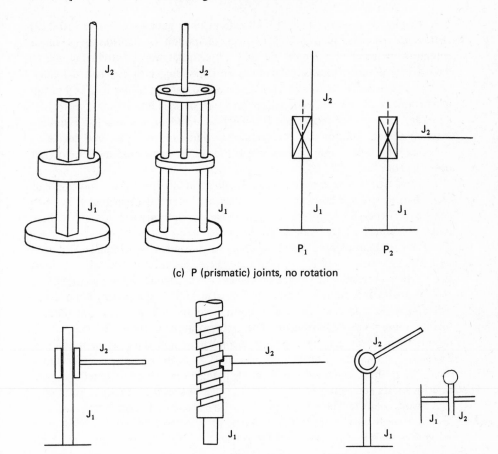

(c) P (prismatic) joints, no rotation

Cylindrical = R + P   H (helical) joint   Spherical = 3R   Flat planar
Symbol $\widetilde{R}\widetilde{P}$ or $\widetilde{R}\widetilde{R}\widetilde{R}$ = $\widetilde{P}\widetilde{P}\widetilde{R}$
Coaxial

(d) Other possible joints

Fig. 5.10 – Types of mechanical joints for arms.

rotation, or (3) by pulleys or gears from motors in the body of the system. They
can also be operated by hydraulic or pneumatic linear cylinders providing
leverage to the second joint in a manner analogous to the human muscle system,
but this gives variable leverage and limited movement.

The second movement which is commonly used is a linear displacement
without rotation which, following Roth[†], we shall call *prismatic* and denote by

---

[†]B. Roth, J. Rastegar and V. Scheinman, 'On the design of computer controlled manipulators',
1st CISM-IFToMM Symposium, Vol. 1, p. 93.

P. This may be powered by (1) hydraulic or pneumatic cylinders directly, (2) motors driving screw mechanisms, (3) rack and pinion mechanisms, or (4) linear actuators. It is called prismatic because the simplest way of producing it is to have a bar or prism of triangular cross-section sliding into a matched holder. Other ways of avoiding rotation are to have two parallel round bars or a piston in a cylinder with one or two parallel round bars outside it.

The third type of movement is called *helical* and corresponds to a simultaneous combined rotation and translation. It thus only provides one degree of freedom. The most common example is the rising hinges used on doors to lift over carpets.

The fourth type of movement is the *cylindrical* movement which corresponds to a coaxial rotation and translation which are independent so that it has two degrees of freedom of movement. It is, however, equivalent to RP and if these two movements have the same axis — this is denoted by $\widehat{RP}$.

The fifth kind of movement is the *spherical* movement which can be produced by a ball and socket and in the most general case has three concentric rotations, two of them corresponding to the movement of the second rod in two directions at right angles and are thus limited to less than 180° if the system is made as a ball and socket because the socket must hold the ball in place and thus be slightly more than a hemisphere. The third rotation is about the axis of the inner sphere which goes through the second joint and thus can be unlimited. The figure shows also a mechanism described by Roth whereby a larger movement about three concentric axles at right-angles can be constructed by using three R joints connected by three brass cranks but so that the three rotational axes always coincide. The spherical system with ball and socket is almost impossible to drive mechanically while the proposal of Roth can be worked with three revolute actuators. It is not necessary to have a separate symbol for the spherical one as it can be represented as 3R or $\widehat{RRR}$.

The sixth movement is the movement represented by two flat planes held against each other, as when one moves a book about on a table; but it corresponds exactly to two P movements in the plane of the table and an R movement about an axis at right-angles.

Thus we can describe all practical mechanical arms by using the R and P symbols with the brace to represent the fact that there is zero length of link between two successive movements. There are other possibilities such as the use of arms or fingers made up of inflatable bags or bellows (e.g. made of glass fibre-reinforced polyurethane) where the application of air pressure simultaneously to all the bellows causes the arms to curl in a circular arc of decreasing radius.

The general purpose of an arm is to place a wrist at a certain point in space with coordinates xyz and then to rotate the hand about the three axes at right-angles to orientate it in a required direction. Thus a mechanical manipulator with six degrees of freedom can in general do the task with no redundancy, that is to say it can only do the given task by putting all its coordinates into one particular set of positions. If it has more than six degrees of freedom it can place the wrist in the required position in an infinite number of ways. If it has fewer than six degrees of freedom it will in general not be able to produce all the three rotations of the hand. A seventh degree of freedom (at least) has to be provided to give the hand the power of grasping, but this has been discussed in the previous section. An analysis of the commercial units available in 1968 by B. Roth[†] showed that the most popular number (six units) was to have five degrees of freedom, one had only three degrees and one had eight degrees of freedom.

There is an important difference between various ways of combining revolute joints, illustrated in Fig. 5.11; one system is that represented by the symbol $\widetilde{RR}$[‡] where there is a very short link between the two and it just consists of two revolute joints at right-angles, in this case the end of the arm can describe a spherical surface. In the second case we have a link between the two of finite length, for example of length equal to the second link, and the second joint can

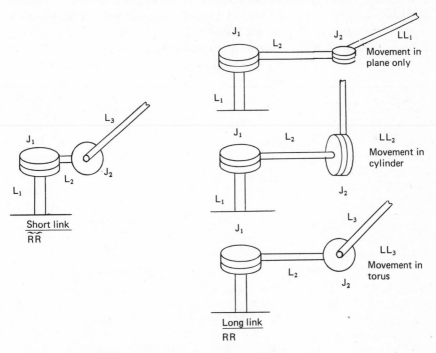

Fig. 5.11 – Arrangements of two revolute joints.

[†]B. Roth, J. Rastegar and V. Scheinman, 'On the design of computer controlled manipulators', *1st CISM-IFToMM Symposium,* Vol. 1, p. 93

[‡]The brace is used to denote the fact that there is a very short link between the bracketed joints.

have its axis parallel to the first one (LL$_1$), or can have its axis along the link (LL$_2$) or can have its axis perpendicular both to the first joint and to the link (LL$_3$). In the first of these cases we have a conventional arm where the end of the second link can move only in one plane and reach any point within a circle (of radius L$_2$ + L$_3$) from the centre outwards; the second link merely provides the variable radius, giving minimum radius when the two links coincide and the maximum radius when they are at 180°. When the axis of the second joint runs along the link L$_2$ (as in LL$_2$) the end of the second joint describes a slightly distorted cylinder of radius equal to the first link and height equal to the second link. In the third case (LL$_3$) the second link describes a complete circle about the length of the first link and this circle can then be rotated about the first joints, so the surface described is a torus of inner radius equal to the excess of length of the first link over the second link and outer radius equal to the sum of the length of the two links.

Roth investigates the design of the most general type of 6R manipulator (Fig. 5.12(a)) with links of finite length between all the joints and shows that this gives something like 32 different ways of reaching a given point while the manipulator corresponding to $\widetilde{RR}\ \widetilde{RR}\ \widetilde{RR}$ gives only four ways of reaching a given point (Fig. 5.12(b)). However, in practice designers have not been concerned

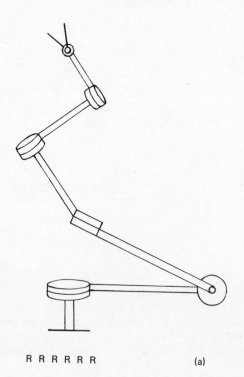

R R R R R R   (a)

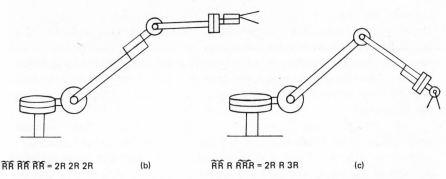

R͡R R͡R R͡R = 2R 2R 2R   (b)   R͡R R R͡R͡R = 2R R 3R   (c)

(roughly equivalent to human arm and hand)

Fig. 5.12 — Arrangements of six revolute joints to give six degrees of freedom of hand location and rotation.

with the complication of being able to reach points in space by different paths and where they have used arms corresponding to 6R they have not usually used more than two links in a manner roughly corresponding to the human arm simplified to R͡R R R͡R͡R (Fig. 5.12(c)), and many of them have used one or more P joints (Fig. 5.13). In general the 6R manipulator is good for a task like

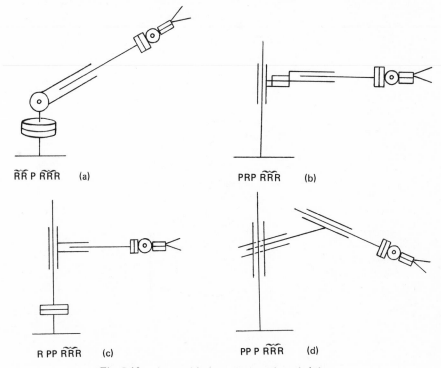

R͡R P R͡R͡R   (a)   PRP R͡R͡R   (b)

R PP R͡R͡R   (c)   PP P R͡R͡R   (d)

Fig. 5.13 — Arms with one or more prismatic joints.

throwing but not for moving along an exact straight line or moving along a chosen path automatically. For precision work the hand must be held close to the fixed end. The revolute manipulators are difficult to counterbalance and have a highly variable rotary inertia especially if the heavy drive mechanisms are placed at the joints. If, however, the drive mechanisms are placed on the body and the drive transmitted by belts, ropes, cables, chains or tapes these effects can be reduced.

The system which is frequently used (Fig. 5.13(c)) of three prismatic, followed by three revolute, joints has the advantage that all the movements correspond exactly to the desired coordinates xyz and the three angles of rotation of the hand. The system can give uniform precision at all points and does not have extreme variations of inertial or gravity effects, especially if it is organised like an overhead crane. This type has been built with only one rotation at the wrist, the hand hanging down from a vertical arm and rotated only about the axis of this arm, in which case it is the four degrees of freedom **PPPR** system and can only pick up objects from above and rotate them about a vertical axis (Fig. 5.14). Figure 5.15 illustrates the volumes which are accessible to some of these arms.

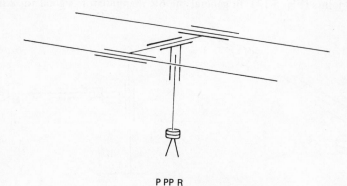

P PP R

Fig. 5.14 – Overhead crane type of robot arm.

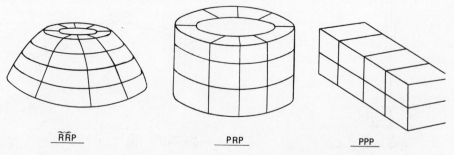

R̃R̂P   PRP   PPP

Fig. 5.15 – Volumes accessible to three types of available robot arms.

Another aspect of design which may be important in some cases is the ability of the mechanical arm to approach the object it is to grip from a variety of different angles. This may be considered by the simpler problem of the two-dimensional arm. Figure 5.16 shows how in two-dimensions an arm with three parallel axis rotary joints and links of equal length can approach a given point P through a whole range of directions between two limiting angles corresponding to the two positions when the first pair of links are in a straight line.

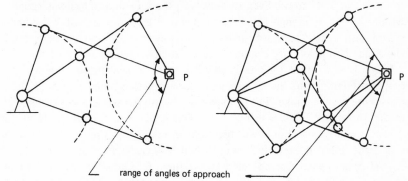

Fig. 5.16 – Range of angles of approach for arm with three equal links in two dimensions.

The human arm as shown in Fig. 5.17 corresponds to an $\overline{RRR}$ (denoted as R_1, R_2, R_3) that is the two shoulder rotations and the rotation about the upper

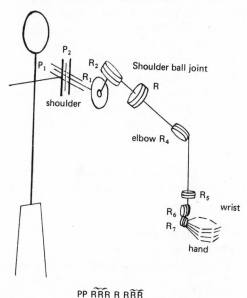

Fig. 5.17 – Mechanical joints of human arm.

arm, all three of which are approximately coaxial, followed by R_4 the elbow rotation which goes from straight to about 15° from fully closed, followed by $\overline{RRR}$ the lower arm rotation (R_5) and the two wrist rotations (R_6 and R_7) which all have axes through a common point. It also has the possibility of raising the shoulder and moving it backwards and forwards in relation to the body (P_1 and P_2) but this need not be regarded as degrees of freedom belonging to the arm since similar movements, together with the sideways movements can be produced by moving the body itself. Even so, however, the human arm has one degree of freedom more than enough to locate the hand in a given place and it is possible when the arm is bent to move the elbow sideways without moving the position of the hand. This possibility can be of great value to reach an awkward place.

· There is one other kinematic possibility for a mechanical arm, which is entirely different from the human arm, corresponding much more closely to the movement of a snake, this is illustrated by the tensor arm constructed by Stanford University[†] shown in Fig. 5.18. This model had six RR joints so that each disc could rotate about two directions in relation to the one before it. By having eight holes in each disc, each of which must be at position 0 or 1, but only one hole in each disc operated at any time, it is possible to put the hand at the end in 8^6 positions and it has been calculated that a 24-link system could reach 0.00004 inches apart in a total volume of 50,000 cu. in. This is a very ingenious design but would clearly be expensive to construct and control because of the enormous number of actuators, valves and cables needed.

Another mechanism for operating a robot arm was used by the author in a preliminary model in the early 60s. This had a single drive motor running a shaft continuously in one direction with the drive transmitted by bevel gears through all rotary joints. Each link could be rotated in either direction by actuating one or the other of a pair of electromagnetic clutches to operate a bevel gear from either side which drove a worm to produce the joint rotation. The system is much simpler than the Lemma[‡] system but does not readily give very accurate position control. Another device used to give precise and rigid positioning was to have an electric motor for the drive operating a back-drivable gear and to use the reverse torque on the starter of the motor to open a spring-operated brake on the joint.

Linear motion joints (P) can have considerable friction when heavily loaded at right-angles to the movement and it is necessary to use some kind of roller bearings.

5.2.2 Operating Mechanisms of Arms
Each joint of an arm whether R or P has to be separately driven to reach a certain value of the corresponding coordinate so that there are three basic

[†]Roth *et al., 1st Symposium,* p. 94.

[‡]L. Kersten, 'The Lemma concept: a new manipulator', *Mechanism and Machine Theory,* 1977, **12,** 77. See section 5.2.2.

problems in the drive (1) providing the power, (2) communicating it to the joint and (3) controlling the exact final position. Other problems are (4) friction and (5) effects of gravity and inertia on the static and dynamic forces. The third of these problems will be described in section 5.3. The power is provided either

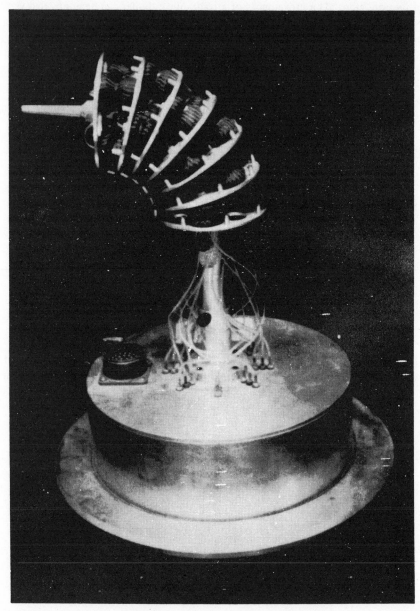

Fig. 5.18 – Tensor arm.

by electric motors or linear actuators or by hydraulic or pneumatic motors, cylinders or rotary actuators (see Fig. 5.19). Electric motors may be geared DC servo motors, which give a rotation angle proportional to the total current supplied or stepping motors which can be operated to give any required number of discrete steps. The movements of a hydraulic cylinder can be controlled very precisely by controlling the total quantity of liquid flowing into it whereas a pneumatic system is always soft and springy.

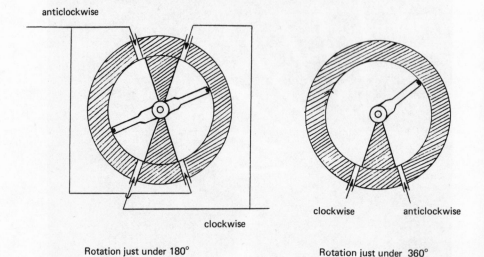

Fig. 5.19 — Rotary hydraulic actuators.

The conventional hydraulic drive valve is shown in Fig. 5.20. A movement of x_1 of the input control arm gives a proportional opening of the spool valve S on

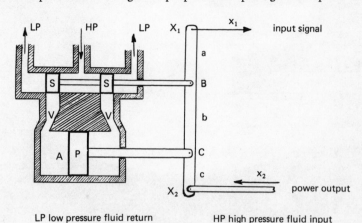

Fig. 5.20 — Simple hydraulic power amplifier.

one side to allow fluid to flow into the main drive piston as long as the displacement x is maintained. This system is not back-drivable in the sense that no amount of pushing on the main piston can open the valve and allow fluid to flow out.

For a small displacement of the spool valve of S to the right corresponding to a positive value of the input signal x_1 a flow of high-pressure fluid into the right-hand side of the main piston P occurs pushing the piston P and the power output to the left by a distance x_2. It is assumed that for small displacemennts of the spool valve

$$\text{displacement } \epsilon = \frac{x_1(b+c) - x_2 a}{a+b+c}$$

the fluid flows into the main cylinder at a rate $= q\,\epsilon$ so that the rate of movement of the piston is

$$\frac{q}{A}\,\epsilon$$

where A is the area of the piston. The leftward displacement of the piston is given geometrically as

$$\frac{x_2(a+b) - x_1 c}{a+b+c}$$

so the transfer function becomes:

$$\frac{x_2}{x_1} = \frac{b+c}{a} \; \frac{1 + \dfrac{c}{b+c}\,\dfrac{A}{q}\,D}{1 + \dfrac{a+b}{a}\,\dfrac{A}{q}\,D}\;.$$

The power output x_2 can be made in the same direction as the signal input by crossing over the connecting ports V. The maximum force on the piston is Ap where p is the maximum hydraulic pressure differential available.

The system reaches a steady state corresponding to a sudden movement x_1 asymptotically with time. The steady state displacement when the spool valve has returned to the zero flow position is given by $x_2 = \dfrac{b+c}{a}\,x_1$.

The arm described by Vertut $et\ al.$[†] uses the motors to act as counterweights to the arm and connects them to the joints by pairs of steel tapes or cables; this reduces the effects of gravity of variable positions but increases the moment of inertia and therefore makes it more difficult to give rapid movements. The

[†]J. Vertut, J. Charles, P. Coiffet and M. Petit, 'Advance of the New MA23 force reflecting manipulator system', *CISM-IFToMM Second Symposium*, 1976, p. 307.

method of counterbalancing used on the Stanford arm[†] is to have a computer control programm which knows the gravity effects on the arm corresponding to the measured values of the coordinates and then fits in a compensating signal to the drive motor so that the control motor is independent of that part of the torque corresponding to the effect of gravity on the arm in that particular position. This does not of course compensate for the load in the hand and so is based on the assumption that the arm is much heavier than the load.

A very interesting arm is described as the Lemma arm,[‡] this is a six degree of freedom arm with all rotary joints (like the human arm it is $\widetilde{RRR}\ \widetilde{RRR}$) using six concentric hollow cylinders so that all the drives could be in the shoulder and carried through the various joints via bevel gears to the wrist and hand.

One very important question in the design of arms is the rigidity. Practical arms are usually designed using hollow cylindrical or rectangular cross-section tubes to give maximum rigidity with light weight since, if the position is measured by means of potentiometers measuring the angular displacement of joints, any bending of the arm links will cause an incorrect position measurement to be fed back. If, however, it is possible to have an independent measurement of the position such as visual sensing then this is not so critical. It is also very important that the rotary joints give very accurate concentric movements and for this purpose large-diameter ball bearings are often used.

For position measurement, by far the most common system is to use rotary or linear potentiometers giving an analogue signal which can be very precise, to one part in 1,000 or better. In the case of hydraulic movement it is, of course, also possible to observe position by metering the volume flow of actuating fluid, provided there are no leaks in the piston.

5.3 DYNAMICS AND CONTROL OF ROBOT ARMS

5.3.1 The transfer function

The human arm is controlled by the three senses, (i) proprioception, that is the sense of movement of the muscles, (ii) touch, either of the object to be picked up or of an obstacle, (iii) visual. It is a very highly sophisticated, fast, accurate and reliable system for reaching, grasping and picking up an object or using a tool for a complex task. The limited range of positioning is overcome by moving the body, walking, kneeling, crawling, standing on a ladder, twisting etc. The limited angular rotation is overcome by the use of cranks, ratchets and tools of all kinds. The problems which we have in designing mechanical arms to reach somewhere near this same sophisticated skill are:

[†]A. Gill, R. Paul and V. Scheinman, 'Computer manipulator control, visual feedback and related problems,' *First CISM-IFToMM Symposium*, Vol. 2, p. 31.

[‡]L. Kersten, 'The Lemma concept, a new manipulator', *Mechanism and Machine Theory*, 1977, **12**, p. 77.

(i)   accuracy of position;
(ii)   control of movement (acceleration, full speed and retardation) so as to arrive at the required spot as rapidly as possible without overshooting or oscillations;
(iii)   sufficient strength to avoid mechanical vibrations of the linkage system;
(iv)   minimum off-set or droop, that is sustained deviation from the desired position due to an external force, such as the weight of an object being held by the hand.

In this section we shall be concerned particularly with the theory of the methods used to solve the second and fourth of these problems.

In the open loop position control system used in robot arms of the first generation, such as the Unimate (see Fig. 5.21), the arm is placed in position by the human controller and carried through the whole sequence of picking up objects or tools, holding them during movement operations. The position of the various joints at the end of each movement are recorded on magnetic tape and the robot is then instructed to repeat these positions by the magnetic tape feeding back the required sequence to the various joints until it is told to stop.

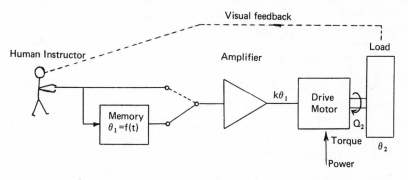

Fig. 5.21 — Open loop position control of a robot (only visual feedback during instruction).

This system is exactly similar to numerical control on a machine tool as it contains no measurement of the actual positions of the joints to verify that they are where the instructions are telling them to be for simple movements. It is 'bang-bang control', that is the controller switches on full power to all the movements until it is near the final desired position and then slows down as fast as possible.

Figure 5.22 illustrates the simple proportional control of position. Information fed back from a transducer such as a potentiometer which measures the exact displacement θ_2 of the joint, is compared with θ_1 the input, that is the required value of θ_2, and the torque or force applied to the load move is proportional to the error $\epsilon = \theta_1 - \theta_2$. In actual practice θ_1 may be changed either suddenly from one value to another or with a sudden acceleration to a constant velocity followed by slowing down. The resulting change in value of θ_2 may be evaluated as follows.

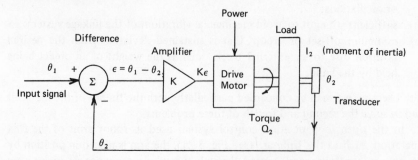

Fig. 5.22 – Closed loop proportional control of a single movement.

The drive motor produces a torque Q_2 which has to overcome the inertial effect of accelerating the arm and object held, after this particular joint which we will suppose has a moment of inertia I_2. It also has to overcome the viscous damping force depending on the velocity $D\theta_2{}^{\dagger}$ of θ_2; thus we can write the dynamic equation for rotor drive

$$Q_2 = I_2 D^2 \theta_2 + CD\theta_2 \qquad (1)$$

The transducer measures the exact value of θ_2 at any given moment and the differencer subtracts θ_2 from θ_1 to produce the error signal $\epsilon = \theta_1 - \theta_2$. This signal is amplified and fed to the drive motor to give the output torque $Q_2 = K\epsilon = K(\theta_1 - \theta_2)$. Equating the value of the torque obtained from this equation with that producing the effect on the load gives

$$I_2 D^2 \theta_2 + CD\theta_2 + K\theta_2 = K\theta_1 \qquad (2)$$

from which we can derive the transfer function

$$\frac{\theta_2}{\theta_1} = \frac{1}{1 + C/K\mathrm{D} + I_2/K\,\mathrm{D}^2} \, . \qquad (3)$$

This may be compared to the transfer equation corresponding to damped harmonic motion

$$\frac{\theta_2}{\theta_1} = \frac{1}{1 + 2d\tau\mathrm{D} + \tau^2\mathrm{D}^2}$$

†D denotes the operator $\dfrac{\partial}{\partial t}$; D² denotes $\dfrac{\partial^2}{\partial t^2}$.

where d is the damping coefficient and τ is the inertial time constant (S) which is the reciprocal of ω_0 the undamped oscillation frequency (S^{-1}). Thus from the comparision we get

$$\tau = \frac{1}{\omega_0} = \sqrt{(I_2/K)}$$

$$d = \frac{C}{2\sqrt{(K I_2)}}$$

As one would expect the inertial time constant τ increases (the response gets slower) as the square root of the inertia and decreases as the square root of the torque amplification K. The damping is provided by the viscous damping of the movement but is reduced by increase of $K I_2$.

Critical damping, occurs when the denominator in equation (3) becomes a perfect square.

$$\frac{\theta_2}{\theta_1} = \frac{1}{(1 + \tau D)^2} = \frac{1}{(1 + \sqrt{(I_2/K)}D)^2}$$

For critical damping $d = 1$ and if C_c is the value of the arm viscous damping factor for this case

$$C_c = \sqrt{(K I_2)} \tag{4}$$

so that critical damping leads to the relation

$$\theta_2 = \theta_1 [1 - e^{-t/\tau}] \tag{5}$$

In an actual control system we do not want critical damping because it will take infinite time to reach the required value of θ_2 corresponding to the imposed value of θ_1. In fact if we suddenly impose a value of θ_1 the solution for θ_2 is given by

$$\theta_2 = \theta_1 [1 - e^{-d t/\tau} \cos (1 - d^2)^{\frac{1}{2}} t/\tau] \tag{6}$$

On the other hand if we make $d = 0$ we have a completely undamped oscillatory system. Thus it is necessary to have an intermediate value of d and the value usually used is $d = 0.6$ to give reasonably fast response with small overshoot. This requires a damping coefficient C given by

$$C = 1.2\sqrt{(K I_2)}. \tag{7}$$

This implies that if I_2 varies (as it does very considerably with an arm with revolute joints or with the grasping of a load) then *the value of C, to give satisfactory performance, must also change in proportion to the square root of I_2.*

This is very important in robot arm design: the human nerve control system makes automatic compensation.

Note, if there is no inertia $I_2 = 0$

$$\frac{\theta_2}{\theta_1} = \frac{1}{1 + (C/K)D} .$$

For a sudden sustained change in θ_1 from zero

$$\theta_2 = \theta_1 [1 - e^{-(K/C)t}]$$

so that θ_2 changes exponentially towards θ_1 with a time constant $\tau' = C/K$. Thus the system can be made very fast by using a large amplification K and a small damping without any problems of oscillation. This shows how *load inertia, e.g. the mass and length of rotating arms, must be kept as small as possible for rapid response.*

Suppose now there is an external torque Q_e on the load resisting the movement: equation (1) changes to

$$Q_2 - Q_e = I_2 D^2 \theta_2 + C D \theta_2 \tag{8}$$

and the transfer function becomes

$$\frac{\theta_2}{\theta_1} = \frac{1 - (Q_e/K\theta_1)}{1 + 2d\,\tau D + \tau^2 D^2} . \tag{9}$$

If we suddenly impose a constant θ_1, the solution of this becomes

$$\theta_2 = \theta_2 [1 - e^{-dt/\tau} \cos(1 - d^2)^{1/2} \, t/\tau] - (Q_e/K) \tag{10}$$

Thus the external torque Q_e produces a permanent error which is proportional to the external torque and inversely proportional to the amplification factor K. This sustained deviation of θ_2 from θ_1 after infinite time is called the offset or droop.

5.3.2 Integral control
Offset or droop can be eliminated so that the final position is exactly that desired (although it may be reached rather slowly) if we add to the torque of the control motor a term proportional to the time integral of the error according to the equation

$$Q_2 = K \left(\epsilon + k \int_0^t \epsilon \, dt \right) \tag{11}$$

In this case the transfer function becomes

$$\frac{\theta_2}{\theta_1} = \frac{[1 + k\mathrm{D}^{-1}] - (Q_e/K\theta_1)}{1 + (C/K)\mathrm{D} + k\mathrm{D}^{-1} + (I_2/K)\mathrm{D}^2} \qquad (12)$$

The term in the denominator by which Q_e is divided becomes very large and thus this offset disturbing effect is steadily eliminated as time becomes great. On the other hand the term in the numerator of θ_1 gives an increased possibility of instability. This type of control is used for the Stanford arm[†] but they integrate the position error only at the end of the arm trajectory and when the position error is within the tolerance set by the measurable error the joint is braked and no longer servoed, to avoid having to use power to hold it in position. It should be noted that the use of integral control will eliminate droop due not only to the weight of the arm itself, in various positions, but also to the load it is carrying; whereas a computer compensated method or the use of counterweights can only eliminate that due to the arm itself.

5.3.3 Derivative control

Figure 5.23 shows the general system for derivative control; the control motor is driven by a signal which is the sum of one term proportional to the error and one term proportional to the rate of change of error $\partial\epsilon/\partial t$, (also called $\dot{\epsilon}$, Dϵ and Sϵ; we shall use Dϵ). This is in principle done by using devices which differentiate, with respect to time, the desired arm joint displacement θ_2. The differentiator may be an appropriate electronic circuit or a small geared electric generator which produces a voltage output proportional to the rate of change of angle. The dynamic equation for arm acceleration is as before (see equation (8))

$$Q_2 = I_2\mathrm{D}^2\theta_2 + C\mathrm{D}\theta_2 + Q_e$$

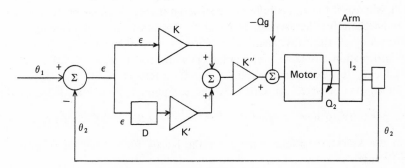

Fig. 5.23 – Derivative control.

[†]A. Gill, R. Paul and V. Scheinman, *1st Symposium,* Vol. 2, p. 31, 'Computer manipulator control, visual feedback and related problems'.

However, the control motor performance equation now becomes

$$Q_2 = K\epsilon + K^1 D\epsilon \tag{13}$$

and the transfer function becomes

$$\frac{\theta_2}{\theta_1} = \frac{[1 + (K^1/K)D] - (Q_e/K\theta_1)}{1 + [(C + K^1)/K]\,D + (I_2 D^2/K)} \ . \tag{14}$$

Note that whereas before the only damping factor was C/K where C was the viscous mechanical damping of the arm we now have an additional damping factor K^1/K which can be controlled by varying the differential amplification. The inertial time constant for zero damping now becomes

$$\tau = \sqrt{(I_2/K)} \tag{15}$$

and the damping coefficient d becomes

$$d = \frac{C + K^1}{2(KI_2)^{\frac{1}{2}}} \tag{16}$$

For the optimum damping $d = 0.6$ and we have the relationship

$$C + K^1 = 1.2\sqrt{KI_2} \ . \tag{17}$$

We can neglect the viscous damping of the arm and put $C = 0$ and produce all the damping we require by controlling K^1.

However, the problem is that the inertia I_2 of the arm succeeding one of the earlier rotational joints, for example at the shoulder, can vary by a factor of 10 to 1 according to the positions of the later joints, even if it is not carrying the load. For example, with an arm like the human one, if the hand is close to the shoulder the moment of inertia for swinging around the shoulder is very small, whereas if the arm is fully outstretched it is very much larger. It is possible to compensate for the varying inertia of the arm itself from a knowledge of the positions of all the subsequent joints once the system has been calibrated and the information stored in a computer. This method is used by Gill *et al.* by introducing the additional amplifier shown dotted at K'' in Fig. 5.23. The amplification factor K'' is varied by the computer from a knowledge of the joint positions to have a value of $K'' = \dfrac{I_2}{I_{20}}$. I_2 is the actual inertia calculated from the known positions while I_{20} is the inertia for a fixed set of position coefficients.

In this case the control motor gives a torque

$$Q_2 = \frac{I_2}{I_{20}} \ (K\epsilon + K^1 D\epsilon)$$

and equating this to the dynamic arm movement equation gives

$$I_2 D^2 \theta_2 + CD\theta_2 + Q_c = I_2/I_{20} (K\epsilon + K^1 D\epsilon)$$

writing $\theta_2 = \theta_1 - \epsilon$ we get the error equation

$$\epsilon = \frac{I_{20} D^2 \theta_1 + (I_{20}/I_2) Q_e}{K (1 + (K^1/K) D + I_{20} D^2)} \tag{18}$$

or the transfer equation

$$\frac{\theta_2}{\theta_1} = \frac{(1 + (K^1/K) D) - (I_{20}/KI_2) (Q_e/\theta_1)}{(1 + (K^1/K)D + (I_{20}/K) D^2)} . \tag{19}$$

Thus the effect is exactly as if the arm had the constant value of the moment of inertia I_{20} except on the external force Q_e. The inertial time constant is given by

$$\tau = \sqrt{(I_{20}/K)} . \tag{20}$$

and the damping coefficient becomes

$$d = \frac{K^1}{2 (KI_{20})^{1/2}} . \tag{21}$$

It follows from (20) that if we take I_{20} as the smallest value of I_2, that is when the arm is contracted as much as possible, then the extra amplifier has always a factor greater than or equal to 1 and the damping coefficient always has a value corresponding to the minimum inertia of the arm, that is one obtains equally good damping at all good positions of the arm. It should be noted, however, that this method does not enable one to compensate for the inertia of the mass grasped at the end of the arm, although it would be possible to feed in to the computer a standard mass figure. In any case the compensation will only be good if the inertia of the arm is considerably greater than that of the object held.

If there were no inertial compensation and d had the value 0.6 at the mean inertia I_{2m} then when the inertia increases to three times the mean, $d = 0.6 (1/3)^{1/2} = 0.35$. So the movement has a large overshoot and oscillates for some time. If the inertia decreases to 1/3 of the mean, $d = 0.6(3)^{1/2} = 1.04$ and the damping is supercritical so that it never reaches the desired point.

Figure 5.23(a) illustrates the way in which the Stanford arm compensates for the gravity torque Q_g on the arm without using counterweights. This is done by having a device which adds into the signal to the control motor, an additional signal to change its output torque by an amount Q_g corresponding to the calculated force of gravity on the arm at the known position. Here again this does not allow for compensation of gravity on the load although it is possible to compensate for an assumed average load.

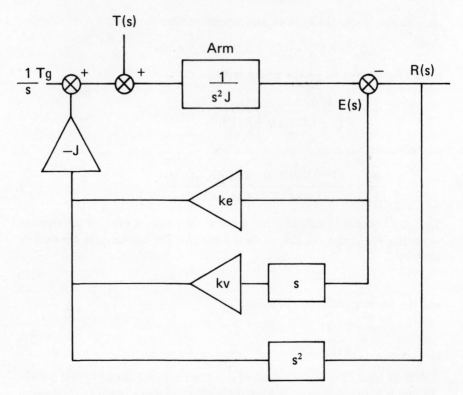

Fig. 5.23(a) – Compensation for gravity (Stanford).

M. S. Konstantinov ('Inertia forces of robots and manipulators', *Mechanism and Machine Theory*, 1977, **12**, 387) has shown that the inertial effects of a general link of a revolute arm may be calculated as though it was replaced by four masses. The equivalent dynamics is obtained if the following three conditions are satisfied.

(1) Mass of the solid link equals the sum of the four contant masses.
(2) The centres of gravity of the solid and of the discrete system coincide.
(3) The inertial ellipsoid of the solid and the discrete system coincide.

This enables the dynamics of such a complex system to be calculated in a much simpler manner since it is only necessary to calculate the behaviour of four point masses.

5.3.4 Bang-bang control
So far we have only considered the case where the torque produced by the control motor is proportional to the error or various functions of the error.

This type of control is not the fastest because it only operates the greatest acceleration when the error is at a maximum and the control torque tends to zero as the error tends to zero. A much faster method of control is called 'bang-bang' control and operates on the principle that as soon as the error exceeds a certain minimum value the control motor is switched on to give its maximum torque or to give the maximum achievable velocity and when the error becomes small it switches off or reverses or a brake is applied. This system is not very good for precision achievement of the final position and is liable to hunt between plus and minus ϵ_0 where ϵ_0 is the minimum error for switching on. 'Bang-bang' control is thus usually combined with a different system of the types considered above when the error comes to a certain minimum value.

5.3.5 The problem of decoupling

Any mechanical arm with revolute joints to achieve the xyz position of the wrist is subject to the fundamental problem that the values of the joint displacement θ_1, θ_2, θ_3, required to provide a given value of xyz are connected by a very complicated matrix relationship, which in general has to be calibrated experimentally, although it can be calculated. The PPP arm system is free from this problem and free from the consequences of variable dynamic behaviour due to variable inertia at different points in the space which can be reached. The ability to control the position xyz by directly activating the joints θ_1, θ_2, θ_3, is called decoupling and is particularly important in the case of telechiric hands where the human controller wishes the slave hand to follow exactly the movement of his hand. The system called 'resolved motion rate control'[†] is one in which the operator grips a control handle and applies forces in the direction in which he wishes to move the slave hand and the rate of movements of the slave hand in the xyz directions is made proportional to these forces by means of a motion resolver which takes in the signal forces in the xyz directions and feeds the appropriate signals to the θ_1, θ_2, θ_3, movements.

Hewit[‡] proposes a different method which is fast and robust and deals with the dynamic decoupling; he calls it 'active force control' and it uses accelerometers on the joints of the arms so that the reaction force vector can be measured. This is subtracted from the known applied force vector to obtain a measure of all the unknown forces such as disturbance torques, friction, etc. The unknown forces are then 'absorbed' by a line of a precomputed constant matrix stored in a memory of reasonable size to leave a set of decoupled linear systems to which conventional control can be applied.

[†]An evaluation of control modes in high gain manipulator systems', D. R. Wilt, D. L. Piefer, A. S. Frank and G. G. Glenn, *Mechanism and Machine Theory,* 1977, **12,** 373.

[‡]J. R. Hewit and J. Padovan, 'Decoupled feedback control of robot and manipulator arms', *3rd Symposium CISM-IFToMM,* p. 251.

5.4 TELECHIRIC ARM CONTROL

5.4.1 The need for force feedback

Telechirics in which the slave arm and hand are remotely controlled by the human master, without force feedback, involve the same control problems as robot arms which have already been dealt with above. The human master issues the instructions for the movements of the arm and hand so that from the point of view of the slave arm these instructions are exactly equivalent to those which would be received from the memory of a robot. Instructing the movements of the slave joint is made very much simpler if the human master moves a control 'hand' which has exactly the same kinematics as the slave hand whether these kinematics are obtained by surrounding the arm of the human in an exoskeleton or by supporting the control hand by separate linkage which duplicates the slave linkage. The other method of driving the control signals is the resolved motion rate control in which the human master operates a gripper on which he can apply the seven forces in the directions corresponding to the seven movements and then the velocity of the slave movement is made proportional to the forces he applies. The problems of variable inertia of the arm and picking up the arm and picking up heavy weights, which have gravity and inertial force, are the same as those with a robot arm for the unilateral telechirs.

Telechirs with force feedback so that the human operator feels a proportion of the force exerted by the slave arm has been shown to have considerable advantages. Wilt *et al.*[†] have compared the carrying out of tasks simulating (1) removing parts from a conveyor and stacking them on a pallet and (2) stowing palletised cargo in a container. They compared 'resolved motion rate control' with 'replica/master control with force feedback'. They found that the mental effort was very much reduced by the replica control and that there was a significant time advantage for moderately complex tasks with close clearances and poor visibility. Bilateral replica control significantly lowered damage-causing movements, but the resolved motion rate control was physically less tiring as the human was not in fact carrying out the movements with his arm.

There are essentially two methods of obtaining force feedback for telechiric systems.

(1) The bilateral system. The master arm and the slave arm are connected essentially symmetrically with position error feedback servo-control both ways and with the force on each arm producing a corresponding force on the other. This force may be 1:1 or it may be scaled up by a predetermined factor so that a small force by the human master produces a very large one on the slave. As the human is pushing the master arm and its drive mechanism, whereas the slave arm is being pushed by the drive mechanism, the force in the link between drive mechanism and arm is reversed. It also follows that the drive mechanism must be back-drivable in each case, that is the arm must be able to move the motor which would not be the case with a worm

drive or a conventional hydraulic valve of the type shown in Fig. 5.20. In that type of valve when there is zero error signal the valve shuts the fluid off completely and the piston cannot be driven. Hydraulic and electric bilateral systems are discussed in sections 5.4.2 and 5.4.3.

(2) Strain gauges or other force measuring devices give signals which can be amplified and used to balance the forces on slave and master arms. In this system the position error signal drives the slave motor and the force difference signal drives the master arm motor so that the human operator feels a corresponding force. This system is discussed in section 5.4.4.

5.4.2 Electrohydraulic bilateral systems

R. S. Mosher has published many papers on his work (e.g. R. S. Mosher, 'Handiman to Handiman', SAE Automative Engineering Congress, Detroit, 9 January 1967, paper 670088). His first system was a directly linked hydromechanical system with a lever force feedback as shown diagrammatically in Fig. 5.24. The first

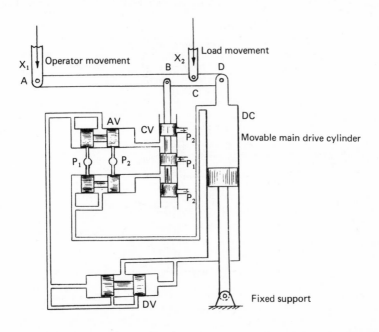

P_1	= high pressure supply	CV	= Control Valve (3 spool)
P_2	= exhaust to reservoir	AV	= Amplifier Valve
		DV	= Damper Valve

Fig. 5.24 – Directly linked hydromechanical force feedback system used by R. Mosher (bilateral).

effect of movement of the operator's arm is to act as a direct lever on the load pivoting about the connection D to the drive cylinder casing. This drive cylinder casing carries fixed to it the three-spool control valve CV, the double cylinder two-spool amplifier valve AV and the spring loaded two-spool damping valve DV. The rotation about D produces a movement of CV which allows the high pressure supply to enter one of the amplifier valves which releases a much greater supply of high pressure fluid to one side of the main drive cylinder. This causes the whole valve assembly to return to the neutral position by which time the load movement X_2 has become equal to the operator movement X_1. The force amplification factor is the leverage AD/CD but movement ratio is 1:1. The spring loaded damping valve moves against the spring one way or the other to reduce the flow rate into or out of the main cylinder.

In order to separate the master and slave so that they are only connected electrically he developed the bilateral electrohydraulic system shown diagrammatically in Fig. 5.26. This requires the use of the back-drivable hydraulic servo valve (Fig. 5.25). The d.c. signal input displaces a reed R by a distance $\pm x_1$ from

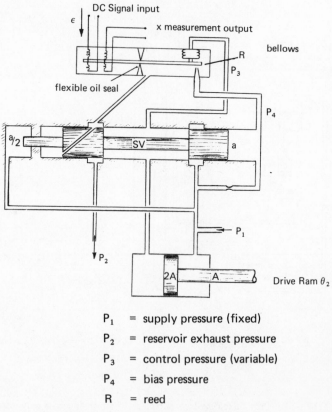

P_1	= supply pressure (fixed)
P_2	= reservoir exhaust pressure
P_3	= control pressure (variable)
P_4	= bias pressure
R	= reed

Fig. 5.25 – Back-drivable hydraulic servo-valve.

the neutral position. This reed works in a chamber filled with oil at the exhaust pressure P_1. On one side it is attached to bellows carrying the controlled intermediate pressure P_c and on the other side it controls the opening of a nozzle discharging fluid at the bias pressure P_b. This bias pressure operates on the large end (area a) of the spool valve V and is controlled by the position of the reed because it is fed from the mains pressure P_s via a restriction. The other end of the spool valve has area $a/2$ and is at pressure P_s so that when the pressure P_b exceeds $P_e/2$ the spool moves to the left. This allows some fluid at P_s to flow to the left of the main drive ram piston. This piston has an area $2A$ but the ram which carries mains pressure P_s has area A, so that when P_c exceeds $P_s/2$ the ram moves to the right. If there is a force opposing the movement of the drive ram P_c will have to rise higher and the bellows will push down the reed against the nozzle causing x_1 to increase and P_b to increase so the spool will open further.

Mosher's bilateral electrohydraulic scheme led to two different types of systems. One of the schemes employed a common error criteria, that is, the error between the positions of the master and slave is made common to the amplifiers driving the servo-valves of the master and slave actuators. The second scheme uses a position servo for the slave and a force feedback for the master, based on the hydraulic pressure changes in the slave actuator.

The common error scheme is illustrated diagrammatically in Fig. 5.26(a). The load held by the slave would increase the control pressure P_c and descend

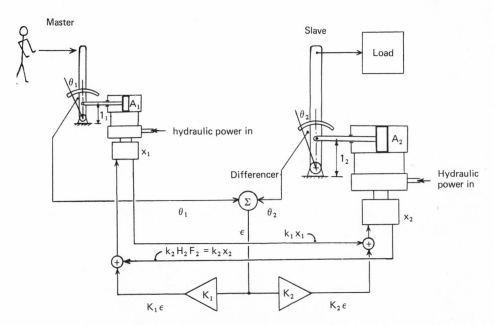

Fig. 5.26(a) — Mosher's bilateral electrohydraulic servo.

subsequently unless the error in positions caused by this generate enough torque in the torque motor to restore the reed in neutral position. At the master end this error would cause the master arm to descend unless counterbalanced by the operator's torque thus providing him a feel of the load held by the slave.

In short, P_c should be counterbalanced by ϵ at the slave end and ϵ should be counterbalanced by P_c at the master end. Thus P_c at the master end has to be modulated by the operator's effort appropriately depending on the dynamics of the slave.

The second scheme is illustrated diagrammatically in Fig. 5.26(b) where the slave servo can be based on the conventional servo-valve of the flow type. However, the master servo valve has to be of the unconventional sort utilizing pressure feedback to provide the back-drivability required to reflect the forces back to the operator. In this scheme the hydraulic pressure changes in the slave actuator are converted electrically for force feedback. That is, a change in force occurring at the slave station produces an electric output which serves as the input of the torque motor in the master station.

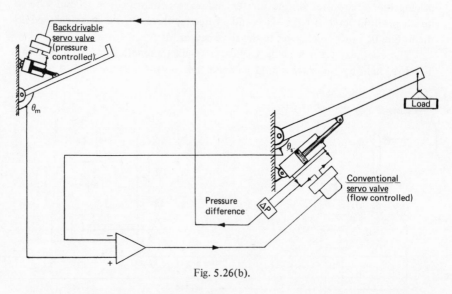

Fig. 5.26(b).

The transfer function for Mosher's electrohydraulic bilateral system can be developed as follows.

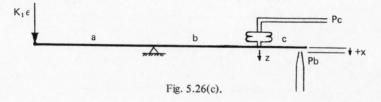

Fig. 5.26(c).

Consider the equilibrium of the reed.

Notation

θ_m	— master arm position
θ_s	— slave arm position
ϵ	— error between positions $(\theta_m - \theta_s)$
x	— deflection of reed against nozzle
Z	— deflection of reed against bellows
k_b	— stiffness of bellows
A_b	— cross-sectional area of bellows
a,b,c	— reed dimensions as shown in figure
P_b	— nozzle back pressure
P_c	— control pressure fed back
P_o	— supply pressure
K_1	— amplification factor of error
σ	— nozzle cross-sectional area
A_p	— cross sectional area of spool

Consider the simpler case where $c \cong 0$ and thus $Z = x$ and $a = b$ (say).
Force balance equation for the bellows is

$$P_c A_b = k_b x \tag{22}$$

The reed's displacement 'x' against the nozzle is determined by the torque balance of the reed. (The nozzle–reed amplifier converts the displacement into a pressure signal and power amplification is carried out by the spool).

A block diagram for the above case can be configured as follows.

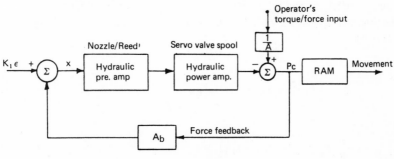

Fig. 5.26(d).

Electrical torque due to position error $(aK_1\epsilon)$ minus opposing hydraulic torque is

$$((P_c A_b - P_b \sigma) b) = (JD^2 + BD + R) \theta \tag{23}$$

where J, B and R are the effective inertia, friction and stiffness terms for the dynamics of the reed, and D is the time differential operator.

Under $a = b$ and $\theta = x/b$ equation (23) becomes

$$K_1 \epsilon - P_c A_b + P_b \sigma = (1/b^2)(JD^2 + BD + R)x \ . \tag{24}$$

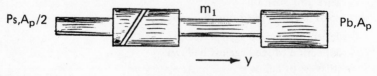

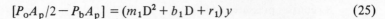

Fig. 5.26(e).

For the spool,

$$[P_o A_p/2 - P_b A_p] = (m_1 D^2 + b_1 D + r_1)y \tag{25}$$

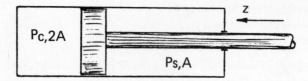

Fig. 5.26(f).

For the ram,

$$(P_o A - P_c \, 2A) = (m_2 D^2 + b_2 D + \gamma_2)Z \ . \tag{26}$$

The parameters m_1, m_2 being the respective moving masses, b_1, b_2 the viscous damping coefficients and γ_1, γ_2 the corresponding equivalent stiffness terms.
Neglecting the dynamics of reed and servo-valve in the low frequency range (i.e. $D \to 0$),

(24) becomes

$$K_1 \epsilon - P_c A_b + P_b \, \sigma = (R/b^2)x \tag{27}$$

(25) becomes

$$A_p \, (P_s \, (1/2) - P_b) = r_1 y \ . \tag{28}$$

When the reed approaches the nozzle, the nozzle back pressure P_b increases (see Fig. 5.26(g)) for a typical nozzle—reed characteristics) and the spool moves to the left thus decreasing the control pressure P_c. This decrease in P_c moves the ram to the left, at the same time allowing the reed to recede from the nozzle. This internal feedback based on pressure only allows the servo valve to be reversible in operation.

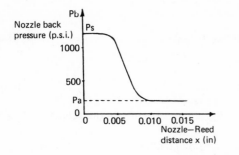

Fig. 5.26(g).

In the linear region of operation, $P_b = K_2 x$ (say)

$$K_1 \epsilon - P_c A_b = Kx \text{ where } K = [(R/b^2) - K_2 \sigma] \tag{29}$$

Let the subscripts m and s denote the master and slave parameters respectively. The above characteristics for a flapper—nozzle arrangement used in the Mosher's back-drivable servo-valve would be as shown below.

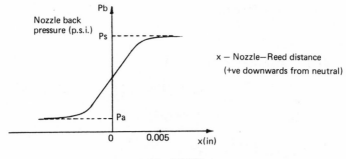

Fig. 5.26(h).

The operator pushes the master arm to create $P_c A_b > K_1 \epsilon$ so that the master reed deflects downwards by x_m (x_m being positive downwards). The equation for x_m (with identical amplifier gains on both master and slave stations) is,

$$P_{cm} A_b - K_1 \epsilon = K_m x_m \ . \tag{29'}$$

Similarly, when θ_s increases against a load, $K_1 \epsilon > P_c A_b$ and the slave's reed deflects upwards by x_s (x being negative upwards) from the neutral position. The equation for x_s is

$$K_1 \epsilon - P_{cs} A_b = - K_s x_s \ . \tag{29''}$$

Consider now the simpler case where the dynamics of the rams are pure inertial.

Then equation (26) becomes as

$$(P_o A_m - P_{cm} 2A_m) = M_m D^2 Z_m \tag{30}$$

and

$$(P_o A_s - P_{cs} 2A_s) = M_s D^2 Z_s \tag{31}$$

$Z_m = h_m \theta_m$ and $Z_s = h_s \theta_s$ where h_m and h_s depends on the geometrical parameters of the articulation.

In equations (29$'$) and (29$''$), usage of identical servo-values would result in $K_m = K_s = K$ (say).

Now, (31) × A_m − (30) × A_s gives

$$(P_{cm} - P_{cs}) 2A_m A_s = h_s M_s A_m D^2 \theta_s - h_m M_m A_s D^2 \theta_m \tag{32}$$

Let us use the equation for the rate of piston movement $A_m DZ_m = q_m x_m$, where q_m is the rate at which fluid flows into the master cylinder per unit displacement of the master reed.

That is, $\quad x_m = (A_m h_m D\theta_m / q_m)$ \hfill (33)

Similarly, $\quad x_s = (A_s h_s D\theta_s / q_s)$ \hfill (34)

Therefore (29$'$)–(29$''$) gives

$$(P_{cm} - P_{cs}) A_b = K (x_m - x_s)$$
$$= K [(A_m h_m D\theta_m / q_m) - (A_s h_s D\theta_s / q_s) \tag{35}$$

substituting for $(P_{cm} - P_{cs})$ in (32) from (35) yields

$$\frac{K}{A_b} 2A_m A_s \left[\frac{A_m h_m D\theta_m}{q_m} - \frac{A_s h_s D\theta_s}{q_s} \right]$$
$$= h_s M_s A_m D^2 \theta_s - h_m M_m A_s D^2 \theta_m$$

$$\frac{\theta_s}{\theta_m} = \frac{D[(2KA_m^2 A_s h_m / A_b q_m) + h_m M_m A_s D]}{D[(2K A_m A_s^2 h_s / A_b q_s) + h_s M_s A_m D]}.$$

With the damped dynamics of the rams (B_m and B_s being the respective damping coefficients) equation (32) will be as follows:

$$(P_{cm} - P_{cs}) 2A_m A_s = (h_s M_s A_m D^2 \theta_s - h_m M_m A_s D^2 \theta_m)$$
$$+ (h_s B_s A_m D\theta_s - h_m B_m A_s D\theta_m) + (h_s R_s A_m \theta_s - h_m R_m A_s \theta_m) \tag{35$'$}$$

Eliminating $(P_{cm} - P_{cs})$ between (35) and (35') will result in,

$$\frac{\theta_s}{\theta_m} = \frac{h_m M_m A_s D^2 + [(2KA_m^2 A_s/A_b q_m) + h_m B_m A_s] D + h_m R_m A_s}{h_s M_s A_m D^2 + [(2KA_m A_s^2/A_b q_s) + h_s B_s A_m] D + h_s R_s A_m}$$

Inspection of the transfer function in above form reveals explicitly the dependence on the ratio A_s/A_m and to a certain extent h_m/h_s.

5.4.3 The electric bilateral telechiric system

Figure 5.27 shows diagrammatically a bilateral system in which the drive mechanisms are reversible a.c. motor/generators. They will be connected to the arms by back-drivable gearing assumed to be of low friction and such that the overall moment of inertia for a movement θ of the arm is I_1 or I_2.

Each motor carries a device measuring θ and one measuring $D\theta$.

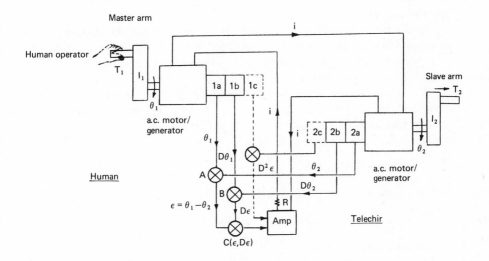

T_1	= input torque applied by and sensed by human operator
T_2	= output torque applied to work (not inertial forces)
θ_1	= rotation angle of input arm
θ_2	= rotation angle of output arm
I_1	= moment of inertia of input arm
I_2	= moment of inertia of output arm + LOAD
i	= current flowing in amplifier circuit, resistance R and both a.c. motor generators
ϵ	= $\theta_1 - \theta_2$ (error)
D	= the operator $\frac{\partial}{\partial t}$

Fig. 5.27 – Electromechanical force feedback bilateral system.

The human applies a torque T_1 to the handle to rotate it by θ_1. T_1 is determined by the feedback force produced by the flow of the current i in the motor/amplifier/resistance circuit, so it has to be calculated.

The devices 1a and 2a send out signals proportional to the angles θ_1 θ_2 respectively, and the devices 1b and 2b send out signals proportional to the rates of change of these angles $(\dot{\theta}_1 \equiv \dfrac{\partial\theta_1}{\partial t} \equiv D\theta_1, \dot{\theta}_2 \equiv \dfrac{\partial\theta_2}{\partial t} \equiv D\theta_2)$. The device marked A receives the signals θ_1 and θ_2 and sends out a signal proportional to their difference $\epsilon = \theta_1 - \theta_2$ — the 'error signal'. Similarly the device marked B receives the signals $D\theta_1$ and $D\theta_2$ and sends out a signal $D\epsilon = D(\theta_1 - \theta_2)$. The device marked C receives the signals ϵ_1 $D\epsilon$ and combines them so that the amplifier F produces a voltage

$$V = K_1\epsilon + K_2 D\epsilon \tag{36}$$

where K_1 is the amplifier proportionality constant and K_2 is the amplifier differential constant.

$K_1 K_2$ can be adjusted, as can the resistance R, to suit the characteristics of the mechanism, but will normally be fixed during a task. However, it would be possible to develop a special feedback mechanism to alter them according to the load or the extension of the arm.

We assume that the two motor/generators have zero electrical resistance (compared to R) but make a back e.m.f. $K_b D\theta_1$, $K_b D\theta_2$ respectively, and produce torques $K_{T_1}i$. We call $\alpha = K_{T_2}/K_{T_1}$ the torque amplification constant. Applying Ohm's law to the circuit we get

$$Ri = V + K_b D(\theta_1 - \theta_2)$$

and hence from (36)

$$i = (K_1\epsilon/R) + [(K_2 + K_b)/R] D\epsilon \tag{37}$$

For convenience call $k = K_2 + K_b$.

The dynamic torque equation for the input shaft, by which the motor resists the human's attempt to accelerate the arm is

$$T_1 - K_{T_1}i = (I_1 D + F_1)D\theta_1 \tag{38}$$

F_1 is the viscous-type arm friction, which we shall regard as negligible.

Similarly for the slave arm

$$T_2 + K_{T_2}i = (I_2 D + F_2) D\theta_2 \tag{39}$$

Here the motor is causing the acceleration. Again we shall neglect the viscous friction, as it is the inertial forces we are concerned with. T_2 can be regarded as a known function of the system — it may be constant or a function of θ_2 — for example in picking up a weight at the end of an arm rotating about a horizontal axis. Note also that I_2 *which is the important inertia of the system* will vary over a wide range both with load in the telechir's hand and with the extension of the arm.

If we eliminate i from equations (37), in equations (38) and (39) we get

$$T_1 - (K_{T_1}/R)\,(kD + K_1)\,(\theta_1 - \theta_2) = I_1 D^2 \theta_1 \tag{40}$$

$$T_2 + (K_{T_2}/R)\,(kD + K_1)\,(\theta_1 - \theta_2) = I_2 D^2 \theta_2 \tag{41}$$

(41) can be rewritten

$$T_2 + (K_{T_2}/R)\,(kD + K_1)\,\theta_1 = (I_2 D^2 + (K_{T_2}K/R)\,D + (K_{T_2}K_1/R)\,\theta_2 \tag{42}$$

This equation shows that θ_2 has no signal from $D^2\theta_1$ so that if at time $t = 0$ $\theta_1 = D\theta_1 = 0$ then θ_2 does not have any acceleration even though we apply a large acceleration $D^2\theta_1$.

Since we are primarily concerned with investigating the effect of the inertia I_2 we can consider the case where T_2 is a known constant. We assume that θ_1 and $D\theta_1$ are known functions of time. Equation (42) is now the standard equation for proportional and derivative control

$$[ID^2 + CD + K]\,\theta_2 = + T_2 + [CD + K]\theta_1 \tag{43}$$

provided T_2 can be regarded as nearly independent of θ_2, as, for example, turning a crank.

In the usual treatment of this equation we use the undamped frequency ω_0 as given by

$$\omega_0{}^2 = \frac{K}{I} = \frac{K_{T_2}K_1}{R\,I_2} \tag{44}$$

and the damping ratio

$$d = \frac{C}{2I\omega_0} = \frac{K_{T_2}k}{2R I_2}\left(\frac{R\,I_2}{K_{T_2}K_1}\right)^{\!\frac{1}{2}}$$

$$= \frac{k}{2}\left(\frac{K_{T_2}}{RI_2 K_1}\right)^{\!\frac{1}{2}} \tag{45}$$

When d reaches the value 1 we have critical damping and the system takes infinite time to reach the equilibrium position if θ_1 is suddenly changed to a new value θ_{10}.

The solution to the differential equation in the case when $\theta_1 \to \theta_{10}$ is

$$\theta_2 = (\theta_{10} + \frac{T_2 R}{K_1 K_{T_2}})\,(1 - e^{-t/\tau}\cos\omega t) \tag{46}$$

where the damping time constant $\tau = \dfrac{2I}{c} = \dfrac{2I_2 R}{K_{T_2} k}$ (47)

and the factor ω, the oscillation frequency (which has dimensions t^{-1}; $\omega t = 2\pi$ corresponding to one oscillation), is given by

$$\omega = (\frac{K}{I} - \frac{c^2}{4I^2})^{1/2} = \left\{ \frac{K_{T_2} K_1}{RI_2} - \frac{KT_2^2 k^2}{4I_2^2 R^2} \right\}^{1/2}$$ (48)

Equation (46) shows that the steady state error due to a load torque T_2 will only be small if $(T_2 R)/(K_1 K_{T_2}) \ll 1$.

If d is very small we have very little damping and the system oscillates for a long time because τ is very large. When $d = 0$, $c = 0$ and the system oscillates according to the equation

$$\theta_2 = (\theta_{10} + \frac{T_2 R}{K_1 K_{T_2}}) \; (1 - \cos \omega_0 t)$$

Thus it is necessary to choose a variable in the control system to give d a value

$$0 \ll d \ll 1 \; .$$

It is usual to choose $d = 0.6$ to give enough damping combined with a reasonably fast response. Suppose therefore we choose $d_0 = 0.6$ for a particular value I_{20} of I_2 and we do this by adjusting the value of R to R_0. Then from (45)

$$R_0 = \frac{k^2}{4d_0^2} \frac{K_{T_2}}{I_{20} K_1} = \frac{(K_2 + K_b)^2}{1.44 \, I_{20}} \frac{K_{T_2}}{K_1} \; .$$ (49)

Thus when the moment of inertia of the arm changes to I_2 due to picking up a weight or lengthening the arm we have a different value of d given by

$$d = d_0 \quad (I_{20}/I_2)$$

if we keep R and the other system factors unchanged.

Thus if $I_2 \leqslant 0.36 \, I_{20}$ the system goes critical, while if

$$I_2 > 4 \, I_{20}$$

it will oscillate with light damping. This sets relatively narrow limits for the moment of inertia variation. If these are to be exceeded it will be necessary either

(1) to carry a compensating 'dumbell' which is dropped whenever the arm is to be extended far or a heavy weight is to be picked up. In this case R must always be small as I_{20} is large and *small R means heavy current and power requirements* or

(2) *to have a range of values of R like a gearbox which can be changed according to the value of I_2.*

It also follows from equation (45) that if we want a large R to reduce the power while d, I_2 are fixed by the external requirements then we reach the *conclusion:*

the factor $\dfrac{k^2 K_{T_2}}{K_1}$ *should be designed to be as large as possible.*

From (42) we can see that the condition for a small steady state error $(\theta_1 - \theta_2)_\infty$

$$\frac{T_2 R}{K_{T_2} K_1} \ll 1$$

Inserting R from (45) this gives

$$\frac{T_2 k^2}{4 d_0^2} \frac{1}{I_{20} K_1^2} \ll 1 \ .$$

These conditions imply

(1) large K_2 — amplifier differential constant;
(2) large K_b — back e.m.f. proportionality of both motors;
(3) large ratio (K_{T_2}/K_1) — many turns on the output motor and/or small amplifier proportionality constant K_1;
(4) $[(K_2 + K_b^2)/K_1] \ll (4 I_2/T_2)$ which is only consistent with conditions (1), (2), (3) if (3) is achieved by large K_{T_2} rather than small K_1, and $T_2 \ll (4I_2/\tau^2)$ (τ given by equation (47)).

Calculation of feedback torque T_1. The second unknown in the system is the feedback torque that the operator finds resisting his attempt to rotate the master arm.

From equation (40) we have

$$T_1 = [I_1 D^2 + \frac{k \, K_{T_1}}{R} D + \frac{K_1 K_{T_1}}{R}] \, \theta_1$$

$$- \frac{K_{T_1}}{R} [kD + K_1] \, \theta_2 \qquad (50)$$

From (42) $\theta_2 = \dfrac{T_2 + (K_{T_2}/R)(kD + K_1) \, \theta_1}{[I_2 D^2 + (K_{T_2}/R)(kD + K_1)]}$ \qquad (51)

but to show the physical effects of the variables we can consider the special case of $\theta_1 \to \theta_{10}$ (fixed) with the solution given in equation (46).

Equation (50) shows that the main feedback torque felt by the operator as a result of accelerations $D^2\theta_1$ and $D^2\theta_2$ is that due to the moment of inertia (I_1) of the control arm. It is probably therefore *desirable to give this arm a moment of inertia I_1 which is proportional to the value of I_{20} for which the resistance R is adjusted at the time,* according to the relation

$$I_1/I_{20} = K_{T_1}/K_{T_2} = 1/\alpha \tag{52}$$

α is the torque amplification factor. This will give the operator a feel for the acceleration $\ddot{\theta}_1$ with force scaled down in the same ratio as T_2 is scaled down. It means that the moment of inertia I_1 must be changed corresponding to each 'gear ratio' R_0, e.g. by sliding a weight along the arm.

If we divide equation (41) by α for the case where (52) holds, and subtract it from equation (40) we get

$$T_1 - (T_2/\alpha) = (I_{20}/\alpha)\, D^2\epsilon + (2K_{T_2}/\alpha R)\,(kD + K_1)\epsilon$$

where ϵ is the error $\theta_1 - \theta_2$.

When ϵ, D_2, $D^2 = 0$ in the steady position the operator feels a torque $T_1 = T_2/\alpha$.

When he inserts an acceleration $D^2\theta_1$ he feels an inertial resistance corresponding to the difference in the acceleration of the master and slave arms. The acceleration of the arm when $D\theta_1$ and $\theta_1 = 0$ is given by

$$D^2\theta_2 = \frac{T_2}{I_2 D^2\theta_1} \qquad \text{(from (42) and (51))}$$

so there is no acceleration if $T_2 = 0$. *Thus a acceleration of θ_1 does not directly produce any acceleration of the slave arm.*

Probably the best results will be achieved if

(1) we have infinitely variable R (slidewire rheostat) and this is automatically adjusted to the optimum value corresponding to I_2 at any moment as a variable speed belt driven cam is automatically adjusted to the optimum drive ratio. The value of I_2 can be sensed by measuring the $D^2\theta_2$ value produced by a measured T_2, and

(2) We introduce an extra differential process into the amplifier so that equation (36) becomes

$$V = K_1\epsilon + K_2 D\epsilon + K_3 D^2\epsilon \;.$$

This will result in an instantaneous response when we have $D^2\theta_1 \neq 0$ and a direct sensing of T_1 as a result of $D^2\theta_2$ and hence of I_2. This will also require differential signals shown as 1c and 2c in Fig. 5.27 and a third differencer G.

Multiple Joints

The above theory has been worked out for a single controlled movement — the rotation θ. Clearly the system of a whole arm with seven degrees of freedom is very much more complicated. However, if we consider an arm (Fig. 5.28) with movements corresponding to two rotations θ and ϕ about the shoulder, an extension x in arm length, and three rotations of the wrist (ζ, η, χ) and one grasping movement δ then there are only two movements to which the above arguments must essentially be applied, viz. θ and ϕ and these can be regarded as two components of a solid angle rotation and controlled by two amplifiers with identical characteristics.

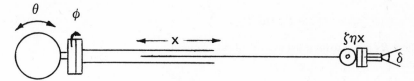

Fig. 5.28 – Multiple-joint arm.

5.4.4 Force measuring servo telechir

Figure 5.29 illustrates the principle of a telechir with the master arm A_1 connected to a motor M_1 which is driven by the signal i_1 produced from the difference of the torques (T_2 and T_1) measured by strain gauges or similar transducers on the drive shafts between the motors and arms. In a hydraulic system the pressure in the drive cylinders can be used similarly. In this case the motors do not need to be back-drivable since if the master applies a torque to the control handle A_1 this gives a signal which allows the motor M_1 to move.

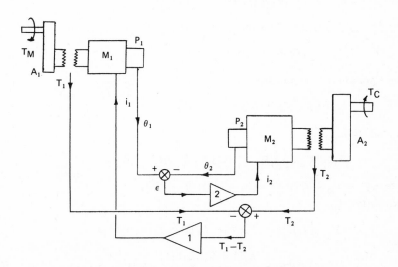

Fig. 5.29 – Electric telechir system with torque feedback.

The motor M_2 is driven in the usual way by the error signal $\epsilon_1 = \theta_1 - \theta_2$. Thus at the slave end

$$K_{T_2} i_2 = T_2 = I_2 D^2 \theta_2 + T_e \tag{53}$$

where T_2 is the measured torque on the shaft (shown broken for clarity) between the motor and arm and T_e is the external load torque (e.g. gravity) while I_2 is the moment of inertia of arm and load.

Now the controller equation gives

$$i_2 = (K_2 \epsilon + K_2^1 D\epsilon)/R_2 \tag{54}$$

where R_2 is the total resistance of motor 2 circuit.
Hence

$$I_2 D^2 \theta_2 + T_e = (K_{T_2}/R_2)(K_2 + K_2^1 D)\,\epsilon$$

If we write

$$k_2 \equiv (K_{T_2} K_2)/R_2 \,,\; k_2^1 \equiv (K_{T_2} K_2^1)/R_2$$

we get

$$I_2 D^2 \theta_2 + T_e = (k_2 + k_2^1 D)(\theta_1 - \theta_2)$$

and the transfer equation is

$$\frac{\theta_2}{\theta_1} = \frac{1 + (k_2^1/k_2)\, D + \; T_e/k_2 \theta_1}{1 + (k_2^1/k_2)\, D + (I_2/k_2)\, D^2} \;. \tag{55}$$

The dynamic equation at the master end is

$$T_M - I_1 D^2 \theta_1 = T_1 = K_{T_1} i_1$$

The equation for the motor M_1 is

$$i_1 = [(K_1 + K_1^1 D)/R]\,(T_2 - T_1)$$

so if we write

$$k_1 \equiv (K_1 K_{T_1})/R, \;\; k_1' \equiv (K^1_T K^1_{T_1})/R$$

we get $T_1 = T_M - I_1 D^2 \theta_1 = (k_1 + k_1^1 D)(T_2 - T_1)$.
Therefore

$$(T_M - I_1 D^2 \theta_1)(1 + k_1 + k_1^1 D) = T_2(k_1 + k_1^1 D)$$

$$T_M = I_1 D^2 \theta_1 + \frac{k_1(1 + (k_1^1/k_1)\, D)}{1 + k_1 + k_1^1 D}\,(I_2 D^2 \theta_2 + T_c) \;. \tag{56}$$

Thus the torque the human master has to apply to the control arm is the torque needed to accelerate this arm plus the torque needed to accelerate the slave arm against the inertia of slave arm and load and the external torque multiplied by an amplification factor. If for example we make $k_1 = 1, k_1^1 = 0$ he feels half the slave torque, but if we make $k_1 = 1/9$ he only feels $1/10$ of the slave torque. In general there is no need to introduce a force damping factor k_1^1 because the slave arm can be damped by the position error damping factor k_2^1.

5.5 CONTROL OF ARMS FOR THE HANDICAPPED

Much work has been done on two types of artificial arm, one is the arm to be attached to the body of an amputee and controlled by the residual muscles, e.g. of the shoulder, and the other is a mechanical arm attached to a table which can be controlled by a quadriplegic by mouth, eye or head movement. In each case the basic problem is to control rapidly but without overshoot up to six movements from a limited number of human control signals.

Orloff published in 1966[†] a study on a control system for an artificial arm in which he designed a hydraulically operated, mechanical version of the human arm to be attached to the body of a thalidomide person born without arms but with some fingers. This was operated with conventional hydraulic servo-valves directly connected to an opposed pair of cylinders producing the movement, preferably by steel tapes wound around a drum. Position indication was by means of a cam follower connected by a lever and spring to the second servo-valve, which thus responded to the difference between the pressure of fluid from the first stage valve operated by a finger and the position of the limb observed by the cam follower. He provided six movements from the operation of two fingers, each on a triangular triple control valve. A variable orifice served as a control signal which controlled the quantity of fluid flowing into an output capsule on the servo-valve. Figure 5.30 shows his design for a 14 inch prosthesis.

Freedy *et al.*[‡] have applied the use of a microprocessor to aid an arm prosthesis using myoelectric data from the muscles of the shoulder girdle region related to the missing limb. They were concerned particularly with the most common case of mid-above-elbow amputation where they required to control the elbow bending, wrist up and down (pronation, supination) hand grasp and often humeral rotation, making three of four degrees of freedom. Each movement was operated by a separate high-speed reversing motor with planetary wave generator and harmonic drive. The motors received signals from a d.c. to a.c. converter, and the nine myoelectric input signals passed through an a.c. to

---

[†] 'Control system for artificial arms and hands', G. Orloff, *Trans. Soc. Instrum. Technology*, December 1966, p. 258.

[‡] 'A micro computer aided prosthesis control, A. Freedy, J. Lyman and M. Salomonov, *2nd CISM-IFToMM Symposium*, p. 110.

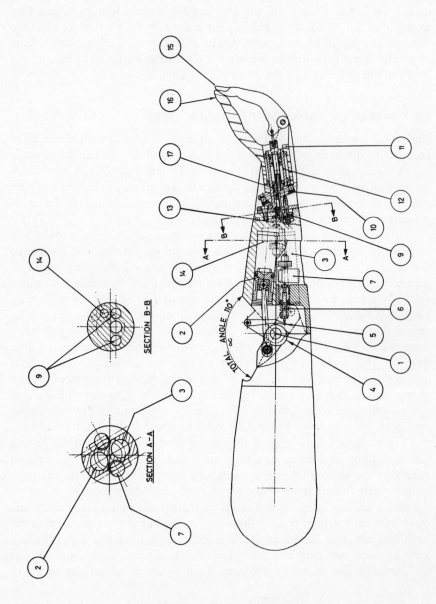

Fig. 5.30 – Orloff's prosthetic arm.

d.c. converter to the digital computer which was calibrated to provide the arm and hand movements desired by the user from his muscular contractions. Calibration was done by user moving his phantom limb and his sound limb in parallel fashion while the computer 'watched' the movement of the latter.

Okada and Kato[†] have studied the shoulder-arm prosthesis for an amputee without the shoulder joint. In this case it is necessary to sacrifice the redundant degrees of freedom of the human shoulder movement because there are not sufficient previously trained muscle groups to control them myoelectrically. They based their system not on end point control, that is on the ultimate desired position of the wrist but on intention control, that is the arbitrary direction vector for the desired movement. Thus the movement is made up of a series of vectors corresponding to intermediate end points and they were able to make the movement much more similar to the natural human movement. Their arm had four degrees of freedom but only positioned the wrist so that one degree of freedom was redundant. A fifth degree of freedom gave forearm rotation and could be used to hold a glass level as it was brought to the mouth. The arm was based on rotary hydraulic actuators and they used signals from the shoulder to control according to the desired movement.

A good example of the manipulator for a tetraplegic is provided by the work of Schmalenbach et al.[‡] They developed a manipulator arm and hand combined with an adapted work place. This has a payload of 50 N, maximum acceleration 5 m/s^2 and maximum velocity 0.6 m/s. Its movements are controlled by a mouth-operated joystick control which could be moved in the xyz directions and had a mouth piece with a central hole which transmits blow or suction to two manometric switches. Two small touch-contact switches are attached to the support of the mouthpiece. One of these switches changes the mode so that the three dimensions of movement of the mouth piece may either position the gripper by movements in the xyz directions or the gripper orientation may be adjusted. The other operates the pneumatic finger. The inputs are analogue signals so that the manipulator carries out the necessary rotary movements to produce the linear movements with velocity corresponding to the degree of movement of the switch. A computer is used to convert from xyz coordinates to joint coordinates and also to enable the hand to maintain its orientation, for example when moving a glass of water. The system enables the tetraplegic to take an object out of a shelf, open a book, turn over pages of the book, using a pneumatic finger which sucks the page, place a sheet of paper in the typewriter and operate a cassette recorder. Most patients have been able to use the device and after training can take a book out of a shelf and open it in 1.5 minutes, turn a page in 10 seconds and dial a four-digit telephone number in 30 seconds. Although much slower than a normal arm these possibilities of action restore a whole dimension of life to such people.

[†] Y. Okada, I. Kato, 'Intention control of the mechanical arm prosthesis', *3rd CISM-IFToMM Symposium*, p. 118.

[‡] *3rd CISM-IFToMM Symposium*, p. 508.

CHAPTER 6

Walking machines

6.1 TYPES OF MACHINES FOR LOCOMOTION

A wheeled vehicle, such as a motor car, with firmly inflated tyres represents the ideal system for minimum energy locomotion on surfaces which are smooth on a macroscopic scale but which have sufficient friction to the wheels to propel and steer the vehicle without slipping. The system has relatively low energy wastage because there are no reciprocating movements in the transmission and propulsion systems and any reciprocating movement wastes power at a rate proportional to the reciprocating mass and the square of the peak velocity unless it is a system oscillating at its natural frequency, such as a pendulum or a mass on a spring which converts the kinetic energy into potential energy, or has a frictionless flywheel. All other walking mechanisms do have reciprocating movements although the waste of power with a leg is minimised if it swings at its natural pendulum frequency during the free forward movement.

The oil resources of the world are sufficiently limited to necessitate a steady rise in the relative price of oil and we shall never obtain a substitute nearly as convenient as liquid hydrocarbons for transport purposes. Thus a minimum energy consumption for any purpose will become increasingly important. For road transport the wheel with the pneumatic tyre must remain the best, although the steel wheel on the rail would be much more economical if trains were built as lightly as buses and lorries with the same duties. However, there are many other areas where the wheel is not the optimum. Four categories are particularly worthy of study.

(i)   Hauling loads (usually a few tons) over soft or irregular ground often with obstacles. The requirement of low energy consumption — which should ideally be roughly that of a lorry on a road — rules out ground-effect machines (hovercraft) and heavier than air machines (aircraft), while airships will probably always have a very high initial cost which will limit their use to very large loads and very long distance or cross-sea trips. The object is to provide a fast 'mechanical elephant'. It is important for use in the Artic, in forestry where it may have to travel over tree stumps up to 0.6 m high, in quarries and mines where fallen rocks are the obstacles, and in marshy and swampy areas in underveloped countries.

(ii)   Agricultural operations require a 'mechanical horse' or bullock which can haul ploughs and cultivators at low speeds with high traction across ground ranging from flooded rice paddy fields to sunbaked rock-hard earth and including soft wet clay and light sandy soils.

(iii)   For movement in situations designed for human legs such as climbing stairs, striding over obstacles or even climbing a ladder. These are needed for fire-fighting devices, factory or domestic robots (e.g. boiler maintenance) devices for carrying heavy weights up stairs, e.g. on stationary escalators or cookers in houses, telechirs for dealing with explosives or other dangerous material in houses, human-propelled stretchers where the patient's weight is on the wheel system.

(iv)   To carry handicapped people or amputees indoors or outdoors, climb kerbs and staircases and give them, as far as possible, normal mobility.

Figure 6.1 illustrates diagrammatically the various types of walking machine that have been studied. The simple rigid wheel (Fig. 6.1(a)) has a constantly changing line of contact with the ground, but a pneumatic tyre has an area of contact such that the tyre pressure balances the weight on the wheel and this whole area is stationary but its boundaries are constantly changing. The pneumatic tyre also absorbs energy due to its hysteresis but does adapt well to small obstacles. When used for high traction on soft ground even with a heavily grooved tyre the wheel spins and grinds its way into the ground.

The tracked system (Fig. 6.1(b)) spreads the load over a much larger area, crawler tractors on farms use 6 lbf/in^2 (4.1 x 10^5 N/m^2)[†] but three times this figure is still acceptable on soft ground and ten times for firm ground. Tracks can be used for ploughing very wet clay but they damage roads and very few farmers can afford the expense of owning one in addition to the general-purpose wheeled tractors. For earth-moving machinery, cranes, etc., the high capital cost is necessarily accepted. Neither a wheel nor a simple tracked system will climb vertical steps or obstacles with a height greater than a small fraction one-eighth to one-sixth of the wheel radius. However, the diamond-shaped track (Fig. 6.1(c)) used on the tanks of World War I can climb much larger obstacles but comes down with a heavy impact on the other side, as does any tracked vehicle at the top of a flight of stairs.

Wheels and tracks with separate feet on hinged rods (Fig. 6.1(d) and (e)) combine the advantages of a substantial bearing area flat on the ground and stationary for a certain period of time as the body moves over them with the fact that separate feet sink into soft ground or gravel and so grip instead of

[†]*'Machines can walk'*, A. C. Hutchinson, *Chartered Mechanical Engineer*, Nov. 1967, p. 480.

piling it up in front. They still do not improve the ability to climb over obstacles, however, and for this it is necessary to come to the devices with separate legs as well as feet. These include the rimless wheel (Fig. 6.1(f)) which can climb stairs of height nearly equal to the radius as can the tracked vehicle with legs held rigidly at right angles to the track (Fig. 6.1(g)). Further advantages are gained by the cam-operated 'square wheel' (Fig. 6.1(h)) which changes the length of the lower legs so that their feet move on a straight line and several can be on the ground at once. Similarly the vehicle (Fig. 6.1(i)) which places its legs vertically down on the ground at the front and picks them up vertically at the back can climb steps of height equal to the radius of the driving wheels.

Finally we come to the systems that have actual mechanical legs — walking machines. Figure 6.1(j), (k), and (l) illustrate three types of automatic legs that have been built in which a single power drive moves the legs through the desired

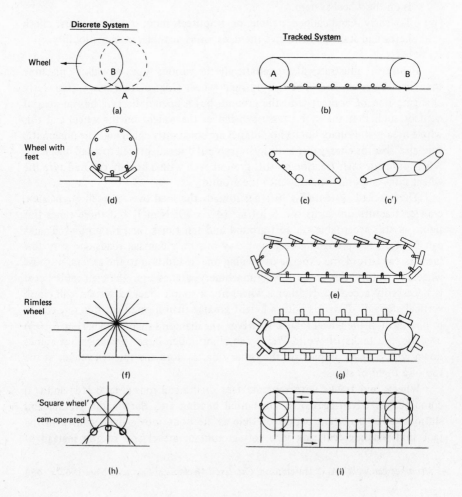

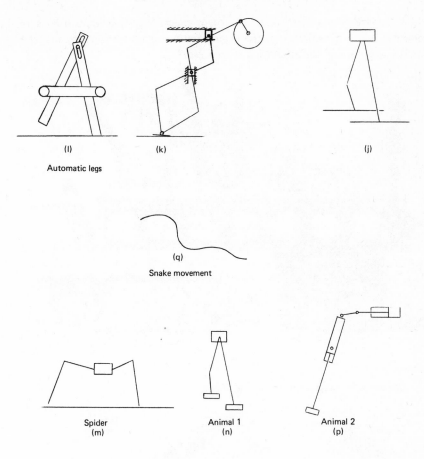

(l)   (k)   (j)

Automatic legs

(q)

Snake movement

Spider   Animal 1   Animal 2
(m)   (n)   (p)

Computer controlled legs

Fig. 6.1 – Walking machines.

path relative to the body by a complex linkage mechanism, so that only the speed of walking can be varied but the leg movement cannot be altered to climb obstacles. Leg systems controlled by computer to cope with obstacles have also been studied extensively in the last few decades, e.g. for moon walking. These allow independent control of the step length and vertical elevation of the foot so that it can climb over obstacles etc. Such machines have had 2, 4, 6 and 8 legs and can have legs like a spider (Fig. 6.1(m)) in which the neutral position of the 'thigh' is horizontal, or like an animal in which the neutral position is vertical (Fig. 6.1(n) and (p)).

   The snake movement (Fig. 6.1(q) has also been studied recently in Japan and is discussed in a special section below.

It is interesting that when the steam engine was first applied to a road vehicle it was not realised that the engine could propel it by driving the wheels and it was given legs for propulsion (Fig. 6.2(a)).

Fig. 6.2(a) – Carriage with legs.

An early fantasy is shown in Fig. 6.2(b)).

Fig. 6.2(b) – Early fantasy walking machine.

There are three main groups of problems connected with walking machines.

(i)   Problems of holding the body reasonably level and ensuring static and dynamic stability during movements, for example up stairs or over ground with large obstacles. These problems are only very partially solved in a wheeled vehicle with rubber tyres and a springing and damping system. Any wheeled vehicle is limited in this respect and, for example, farm tractors frequently overturn when circling too sharply on a slope. A staircase can have an angle of up to $37°$ and stability and uprightness of the passenger and total stability safety are essential in designing a stair-climbing vehicle for the handicapped.

(ii) Steering. Turning a corner is comparatively simple on a wheeled vehicle where the normal system rotates the front wheels by slightly different angles so that the normal to both wheels passes through the same point on the line of the back axle, this point being the turning centre. A longer wheel base requires a greater rotation of the front wheels to obtain the same turning circle radius. It is also possible to design a four-wheeled vehicle in which the back wheels steer by the same angles as the front wheels, in the opposite direction, so that the point of turning lies on a perpendicular line through the mid-point of the vehicle and it is possible to obtain a smaller turning circle. It is necessary to have a differential drive, in both cases, otherwise there is wheel slip because the radius of the circle about which the outer wheel is moving is greater than that of the inner wheel. In a device with legs, steering is obviously an extremely complicated matter and in general movement of the legs has to be controlled by a complex computer program, or by an elaborate cam system for turning. Bessonov and Umnov[†] describe three methods of turning for vehicles with legs (a) discrete turning in which the vehicle stops; the body is supported either by dropping it to the ground, by putting down an extra leg to carry its whole weight or by holding half the legs in a statically stable configuration. This requires a minimum of six legs. The body is rotated through the required angle and then starts off, carried by the remaining legs, in the new direction. (b) and (c) see 6.5.

(iii) The power for walking. It is much more difficult to make a walking system which does not waste power than a wheeled system because in addition to the inefficiency of the engine and transmission it is necessary with a walking system to power a large number of leg joints and much of this power will be wasted in stopping and starting the reciprocating movements. Moreover, it is difficult to avoid vertical oscillations of the massive body, and fluctuations in the forward speed, similar to that felt in a rowing boat, and these both waste power. Theoretically, when progressing along level ground, no power is needed at all and, at the relatively low speeds usually supplied

[†]A. P. Bessonov and N. V. Umnov, 'Features of kinematics of turn of walking vehicles', *3rd CISM-IFToMM Symposium*, p. 87.

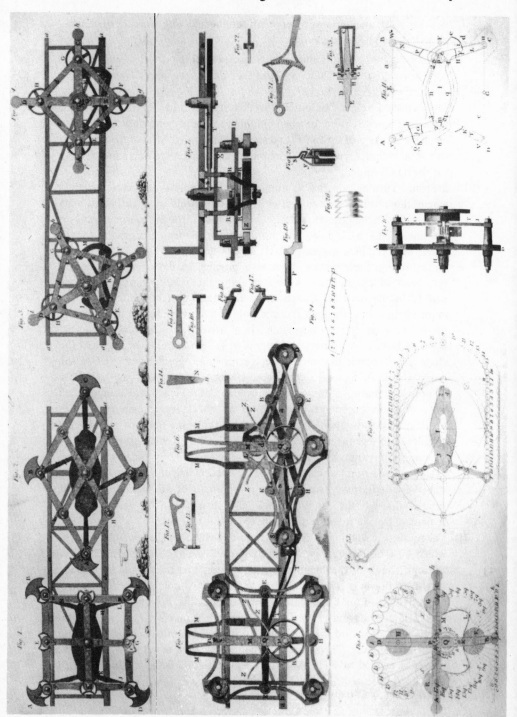

Fig. 6.3 – Gompertz square wheel.

to walking vehicles, air friction is negligible, so that all the power is wasted in mechanical friction in one way or the other.

6.2 MODIFIED WHEELS

Before the caterpillar track system was invented and first used in the tanks for World War I, many people developed modified wheels with feet on them. In 1814 a Mr Lewis Gompertz (A. Jamieson, *A Dictionary of Mechanical Science,* 6th edn, H. Fisher, London, 1828) invented cam mechanisms whereby the rotating of 'square wheels' could move a carriage by placing four feet successively on the ground without vertical oscillation of the axes. Three different mechanisms for achieving the movement are shown in Fig. 6.3. A more practical version the 'Pedrail' (*Harmsworth Encyclopaedia,* vol. 6, p. 4616) shown in Fig. 6.4 was

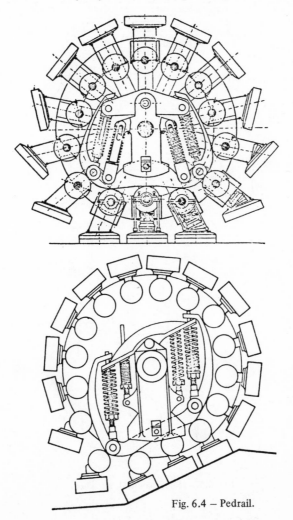

Fig. 6.4 – Pedrail.

used to enable a steam tractor to have a large bearing surface area on soft ground. This also incorporates a 'square wheel' cam system. A very ingenious walking wheel mechanism was invented by the British inventor H. S. Hele-Shaw (US

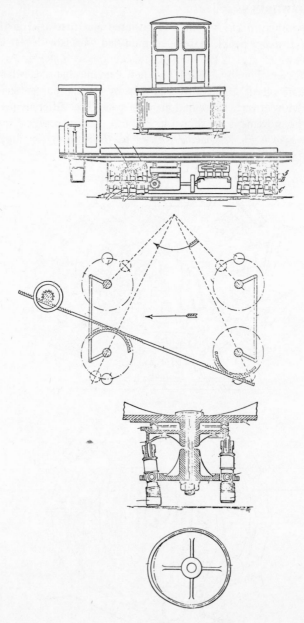

Fig. 6.5 – Hele-Shaw walking wheel.

Patent No. 880526: 1908). The carriage is supported on legs attached to four wheels which have vertical axes (Fig. 6.5(a) and (b). Fixed cams drive the legs downwards to touch the ground at the outside of the circle as the wheels rotate, so that by driving the wheels on the two sides in the opposite direction the carriage is propelled and it can be steered (Fig. 6.5(c) by suitably rotating the four cams to give a differential action; thus a fully four-wheel steering is obtained.

The use of a cluster of 3 or 4 wheels on a circle which can be fixed to hold one wheel or two wheels on the ground when this is level and then allowed to rotate from one wheel to the next to climb on each step (the three-wheeled flip-flop, Fig. 6.6) is an old idea from a Venetian porter's stair-climbing barrow and it has been applied with hand and battery electric power to the stair-climbing wheelchair problem ('Wanted — a stair-climbing wheelchair', US Department of Commerce National Inventors Council, Washington 25, DC, Jan. 1962, design of R. B. McLaughlin).

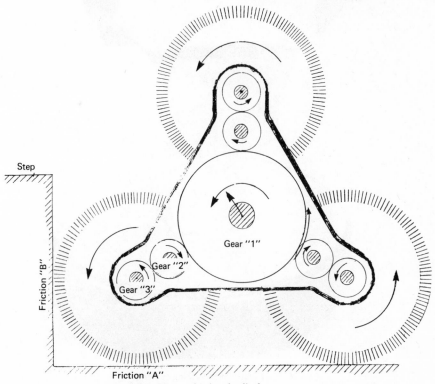

Fig. 6.6 – Three-wheel stair-climber.

For some fifteen years the Mechanical Engineering Department at Queen Mary College studied the possibility of increasing the ability of a powered or hand-operated wheelchair to manoeuvre in cramped spaces. The first attempt

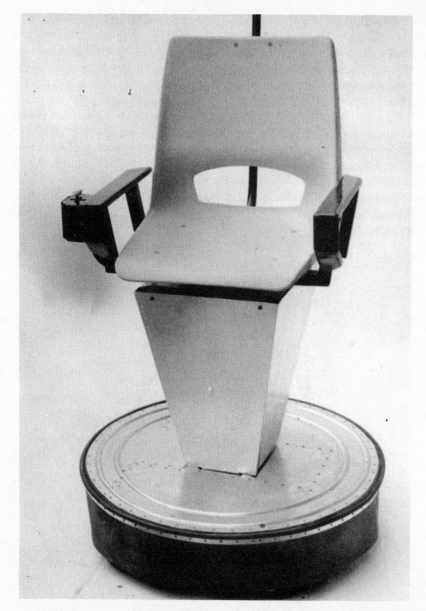

Fig. 6.7 – Eight-wheel walking system.

(Fig. 6.7) was to have two sets of 4 wheels at right angles on separate frames so that by raising the auxiliary frame clear of the ground the main frame could propel the carriage backwards or forwards. By forcing down the auxiliary frame to lift the main frame wheels off the ground the machine could move sideways to left or right. This was intended for mobility in a kitchen but clearly the inability to move in any intermediate direction was too great a handicap.

Fig. 6.8(a) – Special wheel able to roll freely sideways.

The next attempt was a wheel composed of 6-barrel-shaped hard rubber pads (Fig. 6.8(a)) which could be driven by a d.c. reversing motor to its shaft but could also roll freely in the direction parallel to the axis. Figure 6.8(b) shows a chair propelled by four such wheels which can move in any direction. This was subject to considerable wear on the tyres and K. Sheppard proposed and built a system with two reversibly driven wheels with separate motors at the sides and a castor at the front and back (see Fig. 6.9(a)) so that the four wheels form a square with a diagonal pointing forwards. By running one motor forwards and the other in reverse this could spin about its axis and it could move in a curve in any direction by varying the speeds and directions of the two motors. This was

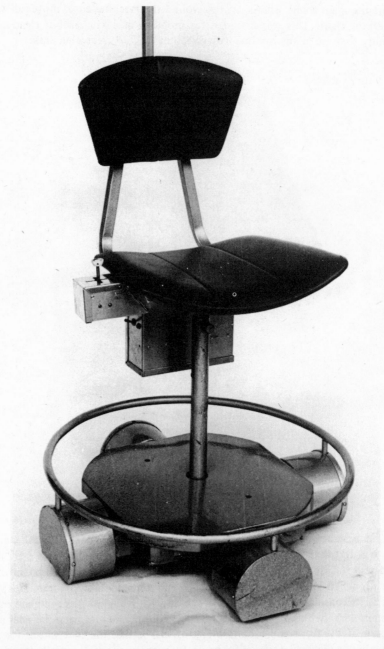

Fig. 6.8(b) – Chair with four wheels of the type shown in Fig. 6.8(a).

Fig. 6.9(a)

done so that a joystick only had to be pushed in the desired direction. It has been built in sizes suitable for children (Fig. 6.9(b)) and adults (Fig. 6.9(c)).

This system had still one great disadvantage — when it came to a shallow step, e.g. the edge of a thick carpet, or to a change of slope the driven wheels were lifted clear of the ground because of the fact that they were in a plane with the front and rear castors. E. Booth introduced the ingenious modification shown in Fig. 6.10, which can also be applied to hand-operated wheelchairs. This enables them to have the powered wheels under the centre of gravity with one or two castors at the front and back and still maintain the powered wheels carrying their load on the ground when the front castors are raised out of line

Fig. 6.9(b)

by a shallow step or concavity in the ground. These systems involved mounting the main wheels and the rear castors on a bogie connected by horizontal axis bearing to the main frame and the front castors. The proportion of the weight carried on the drive wheels can be varied by varying the leverage of the two bars of the linkage as shown in Fig. 6.10(b). This system enables the front

castors to rise over a step equal to about half their radius when it is electrically
propelled. When the hand-operated version is being pushed by an attendant it

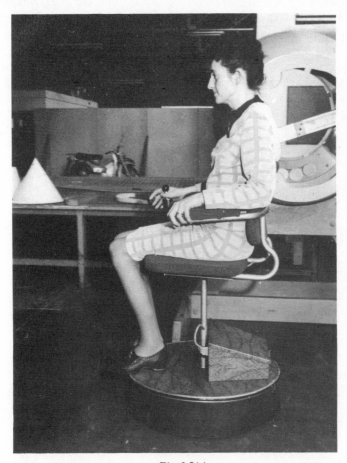

Fig. 6.9(c)

can be raised on to a step of height equal to the height by which the front
castors can be raised against a stop. It is also much easier to push because the
centre of gravity of the person being carried is over the main big wheels and
therefore there is no tendency to slip sideways when on sideways-sloping ground.

Thring has worked on an electrically powered carriage with three or four
rimless wheels to enable handicapped people to climb stairs freely, which essentially places its spokes (rubber-tipped) on the step in front of it so that like a
human foot it has no tendency to roll down the inclined plane formed by the
corners of the step, because the weight is always carried on a horizontal surface.

The motion on the staircase is not perfectly steady because the point of contact of the spokes may be very close to, or even right on, the edge of the step, or it may be up to the distance between the spokes from the edge. However, it is considerably smoother than any system which rolls along the step and then climbs up to the next one, and there is a certain smoothing effect between the front and rear wheels.

Figure 6.11 shows an early working model of this kind of machine with four driven wheels having twelve rubber tipped spokes on each. By having driven front and rear wheels it gradually changes angle as it comes to the beginning and end of the stairs. If the spokes are too sharp they wear the rubber tips

Fig. 6.10(a) — E. Booth's articulated wheel system for invalids.

very fast, but if they are too thick they tend to come on the rounded corner of the step more often and can roll (on the downward movement) or slip (on the upward movement) on to the next step. The optimum number of spokes is also a very important consideration. The wheel should have a radius (when the spoke is loaded) slightly greater than the height of the highest step to be climbed, so that as it rolls forward on the step below the spoke comes forward on to the next step. If there are too many spokes the spoke always begins to lift from a point very close to the edge of the step-tread, and the smoothing effect of the

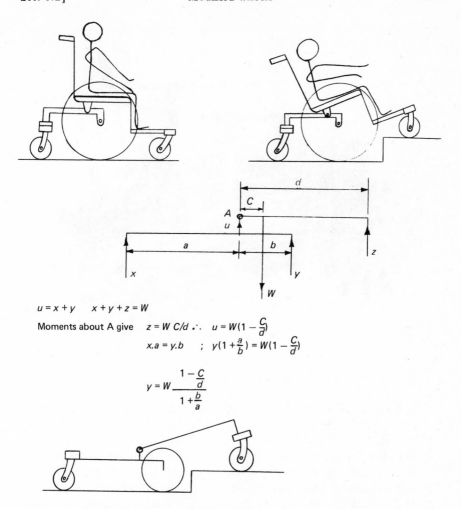

$$u = x + y \qquad x + y + z = W$$

Moments about A give $\quad z = W\,C/d \;\; \therefore \;\; u = W(1 - \frac{C}{d})$

$$x.a = y.b \qquad ; \qquad y(1 + \frac{a}{b}) = W(1 - \frac{C}{d})$$

$$y = W\,\frac{1 - \frac{C}{d}}{1 + \frac{b}{a}}$$

Fig. 6.10(b) — E. Booth's articulated wheel system for invalids.

edge going deep between the treads is lost. If there are too few, the reciprocating motion when going on the level ground at higher speed becomes excessive. Twelve spokes give noticeable oscillation at 3 m.p.h. on the level. The amplitude of the oscillation is inversely proportional to the square of the number of spokes, so 16 or 20 would be better from this point of view though more expensive.

There are two problems which must be overcome in a final design.

(i)   *Stability*. It is desirable that the seat carrying the person shall tilt so that this seat remains approximately horizontal when the chair ascends or

Fig. 6.11 – Invalids stair-climber with four rimless wheels.

descends stairs. If the chair is pivoted about a point above the centre of gravity of the person so that this levelling is automatic on the pendulum principle, then his weight will be thrown in the downstairs direction on the tilting of the chair, making the system fundamentally unstable. It is, therefore, necessary to tilt the seat about a pivot close to the ground or to slide the seat upstairs as the carriage tilts. In either case the weight has to be lifted and thus the levelling mechanism must be powered, unless the movement is carried out by the driver before reaching the stairs at the same time as he changes to low gear for stair-climbing. It is also possible to have front- and rear-wheel drive only when in the climbing mode.

(ii) *Steering.* Steering a system of this type requires front- and rear-wheel drive when it is climbing stairs. Moreover, it should have very good mobility corresponding, if possible, to complete rotation about an axis within the framework. This rules out conventional steering methods unless the driven front wheel can be rotated up to 90° in each direction, a proposal which is being explored at UNAM in Mexico. My own solution is shown in Fig. 6.12. It is preferable to have only three wheels because, when climbing a spiral staircase, which is not a plane surface, a four-wheeled system would require very soft springing to adapt to the surface. By making the front wheel with rollers at the end of all the separate spokes (Fig. 6.13) this wheel will give

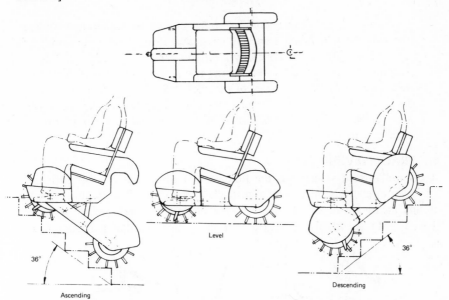

Fig. 6.12 – Stair-climber with three rimless wheels.

Fig. 6.13 – Wheel with rollers.

positive drive for moving forward, but can roll freely sideways. Thus, if the two rear wheels are driven by separate motors, which can be reversed, and the front wheel is driven by a differential gear so that it runs at the arithmetic mean of the speeds of the two rear wheels, it is then possible to rotate about the mid-point of the back axle. The front wheel need not be driven except in the climbing mode and the drives to the rear wheels have to be reduced by a 6:1 ratio mechanical gear in order to give the necessary increased torque.

6.3 TRACKED VEHICLES

A caterpillar-track vehicle has the fundamental advantages of distributing the load over a relatively large area; even so there is a scale effect on this, since the ground area goes up as the square of the linear dimensions and the weight as the cube. In an article, A. C. Hutchinson[†] points out that a 1000-ton tank

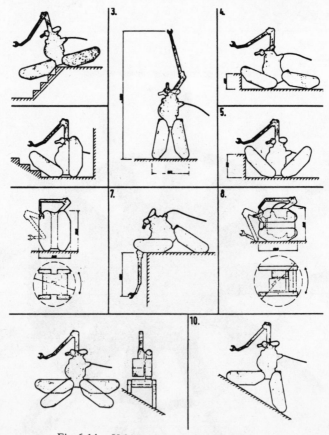

Fig. 6.14 — Vehicle with two sets of tracks.

[†]'Machines can walk', *Chartered Mechanical Engineer,* Nov. 1967, p. 480.

has to have practically the whole ground area taken up with tracks, whereas a geometrically similar 14-ton tank has only a very small fraction, with thin tracks on the edges. One of the fundamental difficulties with the caterpillar track vehicle is that of steering round a sharp radius. While it is theoretically possible to run one track forward and one backward so that the tank rotates around the mid-point, this requires the front and rear ends of the tracks to skid sideways very rapidly. If the weight is equally distributed all along the tracks this produces a very clumsy movement in soft ground and takes up a lot of power, ploughing through the ground sideways. Another disadvantage of the tracked vehicle is that when it moves over a sharp convex change of gound direction, such as that at the top of a flight of stairs, there is a big bang from one position to the other. The tanks of World War I, which had diamond-shaped tracks, could climb quite a large obstacle, but were fundamentally unstable for this reason. The tracked vehicle with two or three tracks on each side, which can be power-orientated at various angles (Fig. 6.14)† can overcome the difficulty of change of ground direction but is necessarily much more complex and expensive. The caterpillar track, like the wheel, is not ideally suited for very soft ground because it has to compact the ground and so do work at the track front all the time it moves. Moreover, it is only possible to have one kind of surface on the caterpillar which, if suited to give good grip on rough ground, does severe damage when taken on roads.

The 'centipede

In the first model (Fig. 6.15(a)) of the centipede the sprung legs were operated with two chains, one arranged half-way up the legs and one attached to the top

Fig. 6.15(a) — Centipede with vertical legs.

† G. Kohler, M. Selig and M. Salaske, *Proc. 24th Conference on Remote Systems Technology*, The American Nuclear Society Inc., 1976.

Fig. 6.15(b) — Centipede with vertical legs.

of the legs, so arranged that the legs were always held vertically. Each leg is separately sprung and can have various types of feet on it (Fig. 6.16). However, a fundamental advantage of separate legs is that if one has a solid rubber pad for each foot, with no track on it at all, it still gives a good grip on soft ground because the front and rear edges of the foot act as the track. The actual weight is taken on a rail with a roller feed to the leg running on it.

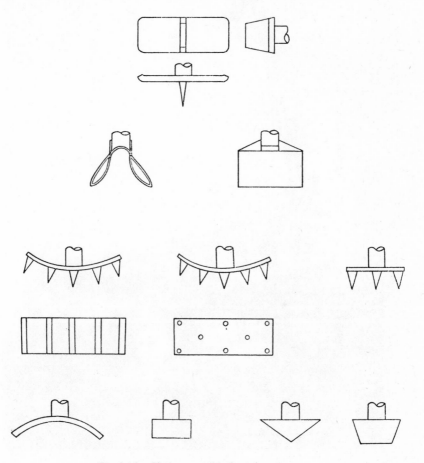

Fig. 6.16 – Various possible foot shapes.

A study of this machine showed that it is not essential to have the legs moving vertically when they come down to the ground and they can come round a circle at the front and still give the same ability to climb stairs. The next version shown in Fig. 6.17 can be described as a caterpillar track with legs. Each element of the caterpillar chain consists of T-shaped piece, joined to the next element by

Fig. 6.17 – Caterpillar track with legs.

rollers at the corners of the crossbar of the T with the stem of the T forming the sprung leg. The two rollers run on rails which are concave upwards so that slightly more weight is taken on the middle feet than on the end ones, to make turning easier. The chains are driven by a hexagonal wheel at each end, with grooves in them that mesh with the rollers. If one has too few corners on these wheels there is too much variation in the speed of the track as the wheel rotates because of the difference in the radii of the circumscribing and inscribing circles of the polygons hence the wheels should be at least hexagons. The rail has to be located with its end at the radius of the circle traversed by the insides of the rollers.

The other proposal (Fig. 6.18) has been specifically put forward for the problems of carrying tree trunks over areas where tree stumps are frequent, and for operating sugar beet or potato-extracting machines in a very wet season. This has a single rubber track supporting low-pressure pneumatic rubber legs, which are preferably elliptical in cross-section, with the long axis in the forward direction, so that they can bend more easily sideways than backwards under load. The belt is driven by a toothed drum on each end, with the teeth meshing with grooves on the inside of the belt. The flat raised part of the teeth on the belt is coated with a low friction plastic and runs between the two drums on a convex-downward smooth steel rail, in the form of a wide plate, which takes the load.

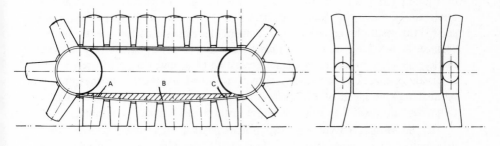

Fig. 6.18 — Walker with pneumatic legs.

6.4 TWO-LEGGED WALKING MACHINES

Some people have used combinations of wheels for stability and legs for movement. Vertut *et al.*,[†] describes a whole series of vehicles with at least two wheels for independent steering and at least two legs or arms to keep the vehicle in a stable orientation by pressing the wheels against the surface. One particular machine can pass through a manhole of 400 mm diameter when the legs are retracted and can then move along the inside of a 800 mm diameter pipe. It can climb a vertical section by pressing the arms hard against the opposite side. Machines are propelled by drive motors on the wheels. Another machine can work on the outside of a pipe and a third can move between equal parallel surfaces, while all the vehicles can move on irregular, near-horizontal surfaces. The work is mainly concerned with supplying the needs of the offshore and the nuclear industries.

[†] J. Vertut, P. Marchal, Y. Corfa and D. Francois, 'Vehicles with wheels and legs. The in-pipe remote inspection vehicle', *3rd CISM-IFToMM Symposium,* p. 476.

In a study of the biomechanical principles of constructing artificial walking systems Gurfinkel and Fomin[‡] used a system consisting of a rectangular cart with castors on each corner and a spider-like leg on the centre of each of the long sides. Each leg has three rotary joints, one fixed to the castor with a vertical axis and two below it with horizontal axes. Each joint was powered with an electric motor. For straight walking both legs were coordinated, but for steering all the elements could be controlled separately.

When we come to two-legged devices without additional stabilisers, however, the dynamically stable design is much more complex.

Two-legged walking, running, jumping and skipping are some of the most sophisticated movements that occur in Nature, because the feet are quite small and the balance at all times has to be dynamic; even standing still requires sophisticated control. If you fall asleep on your feet you fall over. The human stabilises the movement by integrating signals from (1) vision which includes ground position and estimates of the firmness of the ground and the coefficient of friction (2) proprioception, that is knowledge of the positions of all the interacting muscles, the forces on them and the rate of movement of the joints and (3) the vestibular apparatus, the semi-circular canals used for orientation and balance. A very large number of muscles are used in a coordinated way to swing the legs and the muscle is an engine consisting of a power source in series with an elastic connection. The muscle activity is pulsed and there is an energetically optimal walking speed of 3.5 km/hr when the legs swing in a natural oscillation frequency, although they can swing a lot faster without much extra energy.

R. S. Mosher ('Exploring the potential of a quadruped' *International Automative Engineering Congress,* Jan. 1969), describes studies in which the human ability to balance is applied by means of feedback control to enable a man standing at the top of an 18-ft high machine, with the characteristics of an inverted pendulum, to balance it. He found that the learning process was quite rapid because it used the existing balance ability of the man; each subject certainly became transformed from an anxious neophyte to a confident expert in the space of a few minutes. He used his bilateral force feedback system which he calls CAM (Cybernetic Anthropomorphous Machine).

Kato *et al.* have constructed a two-legged robot (Wabot) controlled by a computer which can walk on its large plate-like feet, balance on the foot which is on the ground to give it static stability by leaning outwards at the ankle and also shifting the body in relation to the 'pelvis'. All the joints are rotary and there are 11 joints for walking as shown in Fig. 6.19. The two yaw joints on the 'pelvis' enable the stride of one leg to be in a different direction to that of the leg which is on the ground so that it can turn a corner. Each joint is driven by a hydraulic drive actuator in the form of a direct-acting cylinder, and a rotary

---

[‡]V. S. Gurfinkel and S. V. Fomin, 'Biochemical systems of constructing artificial walking systems, *1st CISM-IFToMM Symposium,* vol. 1, p. 133.

Fig. 6.19 – Kato walking legs.

potentiometer measures the displacement of the joint for feedback information to the controlling computer, (*1st CISM-IFToMM Symposium,* p. 11, 'Information power machine with senses and links', I. Kato, S. Ohteru, H. Kobayashi, K. Shirai and A. Uchiyama; *3rd CISM-IFToMM Symposium,* p. 340, Quasi-dynamic walking of biped walking machine aiming at completion of steady walking, K. Ogo, A. Ganse, I. Kato).

The quasi-dynamic walking of two-legged robots has been taken further by two papers in the 4th CISM-IFToMM Symposium. Quasi-dynamic means that the trajectory of the centre of gravity is in a position of static stability over the foot on the ground at least once during each single walking cycle. In fully dynamic walking the trajectory of the centre of gravity would be in stable equilibrium only when walking has ceased.

The body is regarded as an inverted pendulum, that is as a mass at the top of the light legs. The final model developed at Waseda University[†] has two degrees of freedom (DOF) for each ankle, one DOF for the knee and two DOF for the thigh, making 10 in all. The extra DOF for the ankle and thigh are to give lateral stability by sideways bends. Each joint has a rotary potentiometer and each sole has 4 mini-switches for ground-contact determination. By full computer control they achieved reproducible quasi-dynamic walking with 9 seconds per step and a stride of 45 cm. There is some way to go before it catches up with a fast human walk!

At Osaka University[‡] the dynamic equations of a seven degree of freedom biped walker have been developed to give a hierarchic control system for motion restricted to the sagittal plane, i.e. they do not concern themselves with sideways stability.

Thring has built a purely mechanical, statically stable biped walker, Fig. 6.20 in which each foot is in the form of three sides of a square and it lifts one foot up so that the lateral rods of this foot pass over those of the other, by bending the knee. The centre of gravity of the body is always over the foot which is on the ground.

6.4.1 Studies on human walking

A number of people have studied the way in which a normal human being walks, especially in difficult situations like climbing up stairs. This work is particularly directed at the design of powered and unpowered leg prostheses. A. Cappozzo and T. Leo ('Biomechanics of walking up stairs', *1st CISM-IFToMM Symposium*, vol. 1, p. 115) used a flashing light stroboscope (20 frames per second) to film a person walking up stairs and to photograph oscilloscope screen displays of the horizontal force acting in the plane of progression and the toe and heel components of vertical force in this plane. He determined the locomotion pattern in walking up stairs and regarded the human leg as equivalent to three bars hinged together, each having finite mass corresponding to the thigh, lower leg and foot.

A model of the human walking system was developed by T. Yamashita and H. Yamada ('A study of stability of bipedal locomation', *1st CISM-IFToMM Symposium*, p. 41) to study the dynamic stability. The model consisted of a

---

[†]T. Kato, A. Takanishi, H. Ishikawa and I. Kato, 'The realization of the quasi-dynamic walking of the biped walking machine', *4th CISM-IFToMM Symposium*, p. 408.

[‡]F. Miyazaki and S. Ariomoto, 'A design method of control for biped walking machine', *4th CISM-IFToMM Symposium*, p. 385.

rigid body with mass and 3 moments of inertia and two massless legs. This model confirmed what we know from experience as to the importance of friction between the foot and the ground, for example one has to take much shorter steps when walking on a very slippery surface.

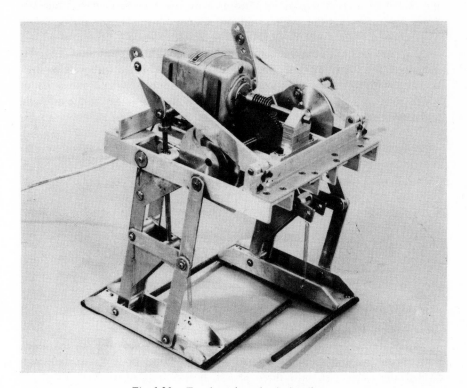

Fig. 6.20 – Two-legged mechanical walker.

A Russian study has been aimed at producing almost normal walking with a sound leg and a powered prosthesis ('Normalisation of walking on prosthesis with an external power source', *1st CISM-IFToMM Symposium*, vol. 1, p. 241).[†] The human walking was regarded as a swinging fork biomechanical system consisting of nine links, one foot carrying the whole weight at any one time with an immovable point of contact on the ground. Each leg is treated as four links, namely, femur, shank, front and rear portion of the foot; the trunk, head and upper extremities were treated as a single link. They studied the movement in in the sagittal plane only, that is they were not concerned with lateral movement.

[†]V. Y. Shishmarev, I. S. Moreinis, A. J. Bogomolov and A. I. Korotkov, Central Research Institute for Prosthetics, Moscow.

A block diagram of the system controlling the walking of a man with one unpowered artificial leg is shown in Fig. 6.21, the dotted rectangle represents the biomechanical linkage consisting of the sound leg (Block T) and the links of the prosthesis (Block P). The upper circuit loop with feedback represents the model of motion regulation performed by the nervous system, while the lower one represents the open automatic system controlling the artificial leg. They applied their work to an above-knee amputee using a prosthesis with a relay control knee unit and external power source. This showed that the over-loading of the sound leg is decreased by 7–10% as compared with walking with traditional type prosthesis (unpowered knee). The optimum condition is that in which there is a minimum asymmetry between the movement of the artificial and natural extremities.

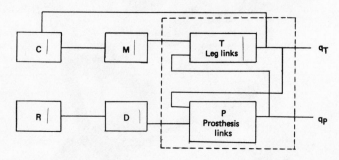

Fig. 6.21 – Block diagram of walking control system for one-leg prosthesis.

In connection with their hydraulically powered biped walking machine, Kato *et al.*[†] have studied the human walking process on a flat surface and divided the single step into three phases, heel-contact phase, kick-up phase and single-support phase.

(1) *Kick-up phase.* When the heel of leg 1 makes contact and the shock is absorbed by synergical movement of knee and angle, the system rotates about the heel of leg 1 and leg 2 kicks up using the toes and then the body is supported solely on foot 1.

(2) *Single-support phase.* The centre of gravity of the body is moved smoothly forward from behind leg 1 to in front of it, by using the kick-up inertial force of leg 2 and the floor reaction force of leg 1. Leg 2 is swung forward and prepared for heel-contact phase.

(3) *Heel-contact phase.* The whole system rotates about the toe of leg 1 and becomes unstable. Leg 2 makes heel contact and system returns to stable

[†]K. Ogo, A. Ganse and I. Kato, 'Quasi-dynamic walking of a biped walking machine aiming at completion of steady walking', *3rd CISM-IFToMM Symposium*, p. 340.

state. It is during this unstable phase that the ankle sweep throw of Judo
is applied to leg 2.

Walking consists of alternate stable and unstable states. However, when designing
a walking robot they maintain a permanently stable walking system as follows.
The floor reaction force, due to friction and normal pressure, can be reduced to
a resultant through a point called the zero moment point, and they ensure that
this point is always maintained within the foot which is on the ground by
moving the trunk sideways, controlled by computer with sensor information.
The kick-up phase is triggered by feedback information of the moment when
the machine starts the double-support phase, following the single-support phase.

The latest work in this field[†] is a joint investigation between Ohio State
University and the Technical University of Warsaw, a wonderful example of
international cooperation in medical engineering. The aim is to produce improved
clinical procedures for diagnosis and management of neuro-musculoskeletal
deficiencies. The motion of a human walker is studied by TV pictures of the
limb positions and ground force plates that give the three components of force.
The human is regarded as a stick figure with seven massive links (body, thighs,
calves and feet), and the dynamic model can be used to calculate the joint
force torques from the observations.

6.4.2 Prosthetic and Orthotic legs

Many laboratories have worked on the problems of providing powered prosthetic
legs to amputees and powered exoskeletons to give movement to people with
paralysed legs. I. Kato, H. Monta and T. Onozuka ('Development of myo-electric
control system for above knee prosthesis', *2nd CISM-IFToMM Symposium*, p. 74)
showed that in walking with a rigid knee prosthesis the body bobs up and down
much more than when the knee can be bent, making walking much more tiring.
By carrying out a computer simulation of the free leg movement in the swing
phase they showed that the same damping coefficient can be used in a prosthesis
with flexible knee walking on the level or down a 10° ramp, but that a larger
damping coefficient is necessary than when walking up a ramp. They devised
a knee control apparatus consisting of a hydraulic cylinder and a spring which
can flex the knee proportional to the knee moment in the stance phase. For
walking up a ramp the control apparatus had a one-way valve to give free movement in the direction of extension but not in the direction of flexion. They are
now investigating the use of the myo-electrical potential of the muscle M. gluteus
medius to control the damping values as this muscle can still give a signal even in
above-knee amputees. The signal from it which has a much higher value in

[†] A. Morecki, S. Koozekanani, R. McGhee, E. Johnson, K. Jaworek, J. Olszewski, S. Rahmani,
'Reduced order dynamic models for computer anslysis of human gait', *4th CISM-IFToMM
Symposium*, p. 430.

climbing up a ramp can be used in conjunction with a foot signal switch which tells when the foot is in contact with the ground and a knee angle measuring potentiometer to control the valves as shown in Fig. 6.22.

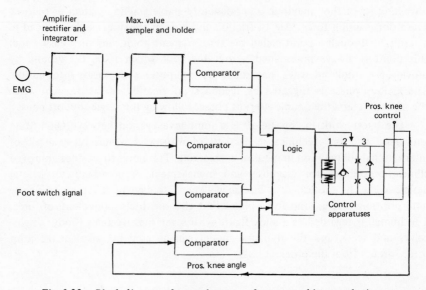

Fig. 6.22 – Block diagram of control system of unpowered leg prosthesis.

The provision of damping resistive torques appropriate to the gait of each amputee joint of an above-knee prosthesis by means of a microcomputer control has been studied at MIT.[†] They used an electronically controlled brake to modulate the damping. A motor/generator rotates a screw with a ball-bearing nut which moves the magnetic particle brake. The screw system gives a 40 to 1 overall gear ratio and a very small amount of electric power is required to control the brake. The user is provided with a set of 10 sliding potentiometers which he can adjust to give the desired damping capacity at various stages in the walking cycle. The next stage will be to provide active power systems to the knee.

M. Vukobratovic[‡] in Belgrade described an exoskeleton for paraplegics (Fig. 6.23) controlled by a computer algorithm which enables the patient to walk and climb stairs. Foot switches are used to vary the program to prevent foot impact.

[†] W. Flowers, D. Rowell, M. Tanquary and H. Cone, 'A microcomputer-controlled artificial joint', *3rd CISM-IFToMM Symposium*, p. 135.

[‡] M. Vukrobratovic, 'Dynamics and control of anthropomorphic active mechanisms', *1st CISM-IFToMM Symposium*, paper 40.

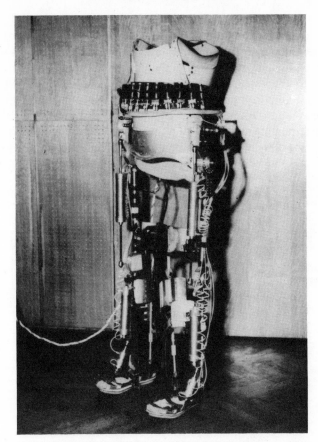

Fig. 6.23 — Active exoskeleton for paraplegics.

The first steps in the development of leg orthoses with hydraulically powered hip and knee movements to enable a paraplegic to walk, stand up and sit down has been described by J. W. Hill[†]. In order to simplify the problem the patient maintains stability not by the control system but by means of two sticks which are also used for steering. Hydraulic actuators have been used because of the large torque (about 12 kg_f.m) required at the joints during sitting and standing. The hydraulic actuators are smaller and lighter than electric motors and can be operated up to 50 bar without danger of explosion. In order to economise on power they allowed the pressure at the output of the pump to fluctuate according to the demand. The block diagram of the control system is shown in Fig. 6.24. The device consists of clothing segments, fabricated from cloth, pressurised

[†] J. W. Hill, 'Hydraulically powered lower limb orthosis', *2nd CISM-IFToMM Symposium*, p. 182.

pneumatic tubes which provide firm contact with the patient and are joined to a metal skeleton of pins to which the power is applied. The mean power required theoretically to drive a human hip has been shown to be 8 watts on a two-second walking cycle. In Hill's experimental set-up they used an electric motor with an efficiency of 65% capable of supplying 100 watts continuously and of giving a very high peak torque. The motor is controlled with a switching regulator using the inductance of the motor armature for the storage. The hydraulic gear pump is only 48% efficient so that the overall efficiency is only 20%. They have developed a walking cycle in which each stage switches itself off and on.

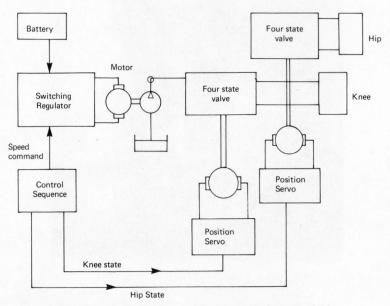

Fig. 6.24 – Control system of hydraulically operated leg prosthesis.

An exoskeletal walking machine for paraplegics has been built at the University of Wisconsin ('Computer control of multi-task skeleton for paraplegics',[†] *2nd CISM-IFToMM Symposium,* p. 233). This uses a 7000 rpm 1/3 h.p., d.c. motor with worm gear and hydraulic speed reduction and cams, rocker arms and coaxial cables to move the leg brace frames in a walking pattern. The primary hip and knee joints are operated by hydraulic rotary actuators, a digital computer controls the system in real time, its outputs are connected to analogue signals which drive the hydraulic actuator control valves through shielded cables. The method of programming the computer is for a normal person to get inside the skeleton and make the required movements with the power turned off so that the results can be stored. It can be programmed for standing up and sitting down,

[†] J. Grundmann, A. Seireg.

walking, stepping over obstacles and stair-climbing; and the user can choose
the program.

To assist people with arthritis of hip, knee or ankle to stand and walk
without the pain due to pressure on the damaged joint, Thring has experimented
with an exoskeleton supporting a bicycle saddle (Fig. 6.25). By means of curved
circular rails under the saddle the upper legs rotate about the hip joint.

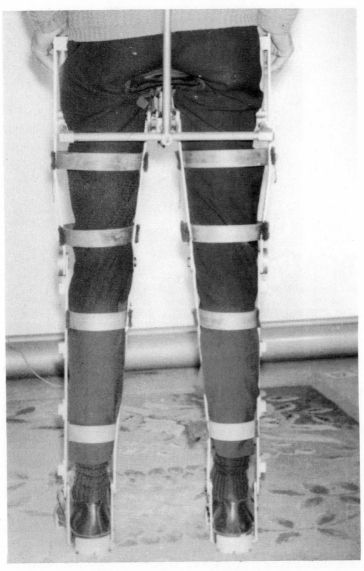

Fig. 6.25 – Exoskeleton for arthritics.

Figure 6.26 shows the linkage mechanism used by Thring in some early studies for a powered orthotic or prosthetic leg. This consists of two parallelograms so that two pneumatic cylinders can provide the whole walking movement and the foot is maintained parallel to the ground. The upper cylinder swings the leg by changing the diagonal of the upper parallelogram, while the lower one lifts the foot by bending the knee.

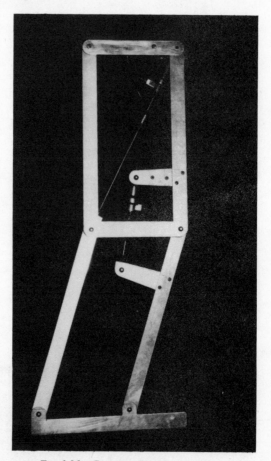

Fig. 6.26 – Pantograph leg mechanism.

6.5 FOUR-LEGGED SYSTEMS

The human two-legged system has only two gaits, *walking* when at least one foot is always on the ground, and *running* when they are both off the ground at once. In both of these stability is dynamic, that is to say if one were frozen at any point one would fall over. The horse on the other hand has the four gaits of walk, trot, canter and gallop.

McGhee and Orin[†] have carried out a computer simulation study in which a digital computer commands the individual joints in response to speed and direction decided by a human controller with a joystick. The *walk* is on a cycle of eight successive phases, with alternately two or three on the ground and operated from 0 to 7 ft/sec. The trot is on a cycle of two phases when alternately the two legs diagonally opposite are on the ground. This is over the range of speed 7 to 31 ft/sec. The gallop is over 31 ft/sec and is on a cycle of eight phases with two, one and no legs on the ground, the no-leg being only one out of eight phases.

Even with four legs there is no possibility of a statically stable walk unless you have wide feet or roll the body sideways as was done in the case of the Japanese two-legged robot (Wabot). A walking machine has, however, been developed in Japan[‡] which has a four-legged crawl, that is to say it has static stability at all stages. Each foot consists of a bar perpendicular to the direction of movement; three feet are always on the ground and the centre of gravity is always within the area of the wide feet on the ground. Each foot moves forward off the ground at three times the speed of the power stroke and the sequence of foot movements is left fore foot, right hind foot, right fore foot, left hind foot. The feet can be raised for the return movement by units of one, two or three according to the unevenness of the ground. The mechanism is shown in Fig. 6.27.

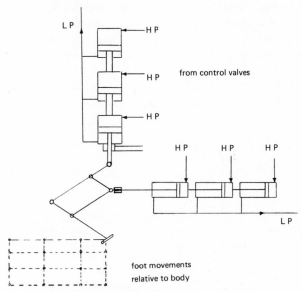

Fig. 6.27 – Spider-type leg-operating mechanisms.

[†]R. B. McGhee and D. E. Orin, 'An interactive control system for a quadruped robot', *1st CISM-IFToMM Symposium*, vol. 1, p. 25.

[‡]K. Taguchi, K. Ikeda and S. Matsumoto, 'Four-legged walking machine', *2nd CISM-IFToMM Symposium*, p. 162.

Each fluid cylinder is digital, i.e. it moves quickly from one end to the other when pressure is applied. The three cylinders at the top provide the raising and lowering of the foot according to how many of them are moved. The three cylinders at the side provide the backwards and forwards movement of the foot. In the power stroke they work sequentially but for the fast return stroke they work simultaneously.

A Russian paper [†] discusses the difficult problem of the kinematics of turn for a four-legged walking mechanism. They are concerned with spider-type legs so that the body can be dropped to the ground for a discrete turning; in this case the vehicle must stop to make a turn. A discrete turn can also be made if a fifth leg is pushed down and the body rotates around this or if there are two groups of legs each statically stable, either by having four legs with wide feet or six legs. However, the more difficult problem of non-stop turning in which the mechanism moves round a curve from one straight line path to another requires a computer or a special mechanism to generate the leg movement arcs of the required curvature. There are two types of turn, one called the *kinematic precise turn* in which the body is always moving in the direction in which it points and the legs slip at the points where it changes from straight line to circular arc motion. The other type is called *kinematic precise circulation* which avoids leg slippage and every part of the body circumscribes concentric arcs of a circle. These two methods of turning may be roughly compared with a four-wheeled vehicle, the kinematic precise turn corresponds to a situation in which the front wheels and rear wheels both steer by equal angles in opposite directions while the kinematic precise circulation corresponds to a car in which only the front wheels steer.

A four-legged spider-type walking vehicle has also been described by Hirose and Umetami.[‡] Two methods of operating the legs are shown in Fig. 6.28. This

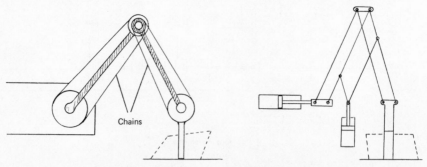

Chains

Fig. 6.28 – Spider type leg operating mechanisms.

[†] A. P. Bessonov and N. V. Umnov, 'Features of kinematics of turn of walking vehicles', *3rd CISM-IFToMM Symposium*, p. 87.

[‡] S. Hirose and Y. Umetami, 'Some considerations on a feasible walking mechanism as a terrain vehicle', *3rd CISM-IFToMM Symposium*, p. 357.

weighs 14 kg and has legs 1.5 m long, giving it the possibility of stride over high obstacles and moving evenly over very rough terrain but is subject to the usual walking problems of poor payload and difficulty of coordinating the joints. They calculate the energy spent on support actuators and swing actuators assuming that the actuators are equivalent to animal muscles in that they absorb energy both in accelerating and retarding and conclude that the spider-type requires less energy to swing the legs than the horse or human type, because the knee-operating muscles of the latter absorb a lot of energy. Mosher[†] has built a walking truck with four legs of similar construction to the human leg, the front two are operated by the driver's arms and the rear two by his legs, the legs are 2 m long and each foot can exert a force of 1500 lb in any direction, the total load being carried $1\frac{1}{2}$ tons. This is operated by hydraulic servos and no electronics. These servos give force feedback to the human operator with a 120-fold reduction.

Another ingenious walking leg mechanism is shown in Fig. 6.29. This was put forward by J. E. Shigley[‡] and was considered in work carried out in Britain for a possible four-legged tank during World War II.[§] The hydraulic leg mechanism

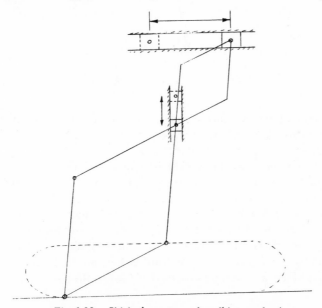

Fig. 6.29 — Shigley's pantograph walking mechanism.

[†] R. S. Mosher, 'Exploring the potential of a quadruped', *SAE Int. Automative Engineering Congress,* Jan. 1969, paper 690191; R. A. Liston and R. S. Mosher, 'A versatile walking truck', *Transportation Engineering Congress,* Oct. 1968.

[‡] J. E. Shigley, 'The mechanics of walking machines,' *1st Int. Conf. on the Mechanics of Soil Vehicle Systems, Turin, Italy,* June 1961.

[§] A. C. Hutchinson, 'Machines can walk', *Chartered Mechanical Engineer,* Nov. 1967.

chosen for the final design is shown in Fig. 6.30. The weight is carried on the roller and the central cylinder only has to operate for raising the leg. Shigley's mechanism can be hydraulic or mechanical.

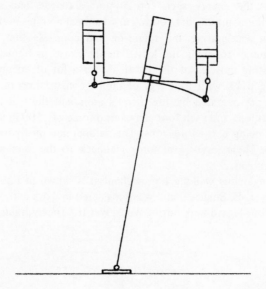

Fig. 6.30 – Hydraulic leg mechanism chosen by A. C. Hutchinson.

Thring has used the fact that the ideal foot movement is close to that of a link of a chain going around two sprockets to produce a very simple walking mechanism for a tractor. A small model is shown in Fig. 6.31. There are two legs on each side and when the weight is transferred from one to the other that side drops by a distance equal to the radius of the sprockets. During the supporting movement the weight is carried on a rail via a roller on the axis of the link from the leg to the chain. Steering can be done by using wheels in front as in the model shown: in an application for low power cultivator for countries where energy is precious it could be used as a 'mechanical bullock' and steered by long handles. For a 'mechanical elephant' or high powered tractor it would have four pairs of legs and steer by varying the speed on the two sides.

6.6 SIX- AND EIGHT-LEGGED WALKING MACHINES

With a six-legged walking machine it is possible to walk with static stability all the time since the body can be supported alternately on three legs forming a triangle and the other three legs forming a reverse triangle and the centre of gravity moving well within both triangles, passing through the centroid of the support triangle half way through the step forward. It can, of course, have many

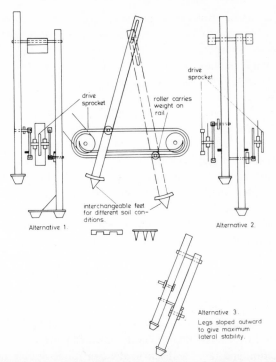

drive
sprocket

drive
sprocket

roller carries
weight on
rail

interchangeable feet
for different soil con-
ditions.

Alternative 1.   Alternative 2.

Alternative 3.
Legs sloped outward
to give maximum
lateral stability.

Fig. 6.31 – Thring's walking tractor. (a) Details of operating mechanism for two
legs on one side. (b) Small model.

other gaits and Bessonov[†] shows how symmetry reduced the number of possible gaits to 24. In a Russian study of six-legged walking machines[‡] a system is modelled with tactile and position sensors on the legs and a distance measuring scanner that gives information on a strip of ground several bodies ahead. The walker has a rectangular body and six equal, symmetrically positioned three-degree-of-freedom legs. The information is put into a digital computer which plans the motion of the body and the legs and then controls the leg action and the environment perception. In this way it is possible to move with minimum energy and disturbance over rough terrain, by memorising the leg position on the ground, measuring the real load on the leg and correcting the leg motion at all phases of the movement for variations in the ground. The system is designed to have two gaits, the tripod pair and a gallop, in which all the legs come off the ground simultaneously. Russian work[§] on a six-legged walker in the 4th Symposium describes an experimental model with various possible gaits which has a three-component force sensor and a surface-contact sensor in each leg and a gyroscopic pendulum on the body to give inclination in two directions. There is also a triangular optical rangefinder to scan the ground in front with two degrees of freedom rotation. In the autonomous mode the machine can bypass large obstacles and climb ledges.

In a paper to the 3rd Symposium [||] McGhee *et al.* describe further work on the walking hexapod with three independently powered joints in each link. Each joint provides position and rate feedback to the control computer and one link is fitted with a vector force sensor. A human operator controls the desired direction and speed but the computer has control software for automatic adapt-ation for uneven ground and decides the actuator torques to minimise the energy consumption. The control system is shown in Fig. 6.32. He shows that it has a real advantage on soft and irregular ground, with potential applications in Arctic transport, forestry, fire-fighting and unmanned extraterrestial exploration. The operator's leg command will only be used on very rough terrain.

In a paper to 4th Symposium[††] the same authors develop a method of carrying out the dynamic calculations by simulating active articulated mechan-isms such as an adjustable link, four-bar linkage design for the legs of a robot

[†] A. P. Bessonov and N. W. Umnov, 'Choice of geometric parameters of walking machines', *2nd CISM-IFToMM Symposium*, p. 62.

[‡] D. E. Okhotsimsky and A. K. Platonov, 'Walker's motion control', *2nd CISM-IFToMM Symposium*, p. 216.

[§] E. A. Devjanin, V. S. Gurfinkel, V. A. Kartashev, A. V. Lensky, A. Y. Shneider and L. G. Shtilman, 'The six-legged robot capable of terrain adaptation', *4th CISM-IFToMM Symposium*, p. 375.

R. B. McGhee, C. S. Chao, V. C. Jaswa, D. E. Orr, 'Real time control of a hexapod vehicle', *3rd CISM-IFToMM Symposium*, p. 323.

[††] R. B. McGhee, C. S. Chao, V. C. Jaswa, D. E. Orr, 'Dynamic computer simulation of robotic mechanisms', *4th CISM-IFToMM Symposium*, p. 337.

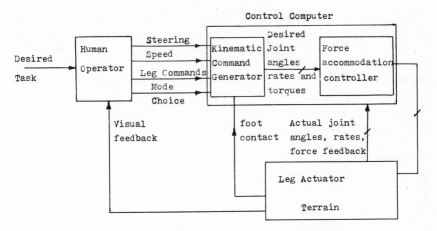

Fig. 6.32 – Man/computer control of hexapod walker.

vehicle (see Fig. 6.33). The foot has two degrees of freedom when in the air but when on the ground the two powered movements must be coordinated so that the body moves parallel to the ground.

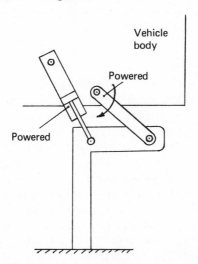

Fig. 6.33 – Adjustable link leg for hexapod vehicle.

6.7 ONE-LEGGED HOPPING MACHINE

A one-legged hopper has only one gait but requires an extremely sophisticated dynamic stabilising system. An experimental machine is being built† primarily

---

†M. M. Raibert, 'Dynamic stability and resonance in a one-legged hopping machine', *4th CISM-IFToMM Symposium*, p. 419.

to assist the understanding of the locomotion problems of balance, resonance and dynamic control without the complication of coordinating different legs. The movement is confined to a plane surface at 45° to the horizontal on which it is supported by air bearings, the 'ground' being a surface at right angles to this. The theory treats the ideal hopping machine as a leg of finite mass (m_1) connected to a body of finite mass (m_2) by a powered rotary joint. The body has a fixed moment of inertia about the joint, but the leg can slide in the joint resisted by a strong spring and this sliding displacement is also powered. The ground is taken as equivalent to a spring and viscous damping. The device hops when the sliding movement is rapidly powered downwards to raise the mass of the body, while the rotary power is used to decide the direction of hopping and to achieve balance. In the actual model the sliding bearing is joined to the body by two pneumatic cylinders forming a V with their piston rods hinged to the bearing so that when they both work together the linear movement is powered, but by operating only one the rotation is produced. The main energy losses occur when the foot strikes the ground and when the spring stretches past its rest length to hit the stop. The return shortening of the leg can occur (1) at lift-off which gives maximum foot clearance for hopping over an obstacles (2) at peak height which maximises time between actuators (3) upon touch-down which minimises ground impact forces. The third system is used in human hopping — bending the knees on landing to minimise shock.

6.8 SNAKE MOVEMENTS

Umetami and Hirose,[†] have carried out a detailed study of the way in which a snake moves and built a number of mechanisms which move in the same way. The snake can wind its way through obstacles and use a fixed peg to push itself forward. It can also go round an obstacle and continue in the same direction. On a flat surface it moves rapidly by passing a sine wave in the lateral direction along its body, but at the same time curving the body so that the peaks of the sine curve are lifted off the surface and the contact points are on the axis. It can also act as a gripper by coiling round an object of any shape. The snake makes use, both of a central nervous system which makes decisions and recognises the environment as a whole and a spinal nervous system which receives tactile sensations of obstacles against the side of the body and controls the muscles for the bending. They constructed a simple snake walking mechanism, the first version of which had ratchet wheels on each section, which could only rotate in one direction. More recent devices are what they call an 'active cord mechanism' which consists of 20 links, each carrying on—off switches representing tactile sensors on each side. This can be used both as a means of propulsion in the midst of obstacles and to grasp objects by coiling round them. This system was

[†]Y. Umetami and S. Hirose, 'Biomechanical study of serpentine locomotion', *1st CISM-IFToMM Symposium,* vol. 1, p. 171.

too elaborate for a gripper and they developed the soft gripper[†] in which the pulling of a single wire moves successive links to apply an equal pressure of each link against the obstacle.

Another type of 'active cord mechanism'[‡] has three-dimensional ability and consists of a series of cylinder segments. Each joint between two of them is rotated by a geared motor controlled by a microprocessor and the joints are alternately perpendicular and oblique. By rotating an oblique joint and the next perpendicular one through an angle up to 180° it can be made to bend by an angle up to twice the angle of obliquity. If it is given powered wheels it can then move on completely uneven territory and cross a considerable crevasse. The wheels are fixed on collars free to rotate about the cylindrical trunk sections so that gravity keeps them at the lowest part as the trunk section rotates. The machine will eventually have visual or tactile sensors giving information to the computer which controls the 3 W d.c. incremental motors which drive the section rotations with a 1/2083 gear reduction.

6.9 COMPARISON OF DIFFERENT METHODS OF PROPULSION

This chapter can be summarised by making a general comparison of the advantages and disadvantages of six different methods of propulsion.

(i)   *Wheels.* These are by far the most efficient and simple for hard, smooth-surfaced roads. By using soft pneumatic tyres they can be made to go over roughnesses small compared with the radius of the wheel, but the energy consumption immediately goes up very considerably.

(ii) *The caterpillar.* This is good for soft ground as it can distribute the load over a large area. As the size of the vehicle goes up the contact area must go up according to the weight (proportional to cube of the vehicle dimensions), while the ground area of the vehicle only goes up as the square, hence the proportion of the ground area covered by the tracks must increase. It eventually reaches unity corresponding to the largest mass that can be carried on a landship. The caterpillar is very much more expensive to construct than the wheel and the power required to drive the mechanism is considerable. It does, however, have the advantage over the wheel that the point of contact with the ground remains fixed for the length of the track and therefore at high traction there is not the same tendency to dig into the ground as is found with wheels, for example in ploughing.

[†]S. Hirose and Y. Umetami, 'The development of soft gripper for the versatile robot hand', *Mech. and Machine Theory,* 1978, **13,** p. 351.

[‡]S. Hirose, S. Oda and Y. Umetami, 'An active cord mechanism with oblique swivel joints and its control', *4th CISM-IFToMM Symposium,* p. 395.

(iii)   *Rimless wheels.* These are excellent for climbing stairs or moving over obstacles of size comparable to the radius of the wheel but they do give a vertical oscillation when running at high speed on level surfaces, which varies inversely as the square of the number of legs. They are reasonably cheap to construct and operate and the energy consumption need be no more than that of wheels.

(iv)   *Legs.* Legs are very good for very high thrust and for convenient changes of the shoes for different types of ground. They are also very good for very irregular surfaces especially if they have variable step-height control by a computer or by human. They can also produce considerably less damage to standing crops than wheels or caterpillars. The chief problem is to construct them simply and cheaply and they will certainly use more energy at high speeds than continuously rotating devices.

Legs involve masses which reciprocate and reciprocation involves a severe loss of energy unless the system is oscillating at its own natural frequency, i.e. almost all the kinetic energy of movement is being converted into potential energy of a coiled spring or a raised mass as in a pendulum. Humans and animals use much less energy when their legs are swinging near their natural pendulum frequency but their muscles are not like reversible springs, they absorb energy for stopping as well as for acceleration. Most mechanical walking systems similarly waste energy in decelerating

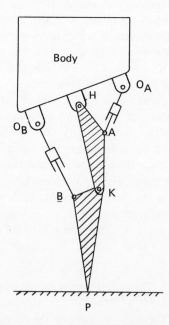

Fig. 6.34 – Low energy mechanical leg.

limb movements. The use of springs (as in the hopper discussed in section 6.7) or regeneratively braked electrical or hydraulic systems can reduce this power very much. Waldron and Kinjel[†] point out that the efficiency of the animal or human leg is improved by the fact that some muscles act across two joints. This is because, if the supporting leg is allowing the body weight to fall during the last part of the support movement, then the power muscles are absorbing energy which must come from elsewhere, i.e. raising the body during the early part of the movement. They propose the mechanism shown in Fig. 6.34 for a mechanical leg to avoid this loss. The four-bar linkage $O_B BKH$ is so proportioned that $O_B B$ can be kept very nearly constant when the body moves forward along a horizontal line and the foot P is fixed on the ground. Thus the actuator B does very little work during the support phase and the actuator A is only providing the work against external resistance.

(v)   *Rams.* For very high thrusts, as in a coal or rock-cutting machine, walking systems based on rams, with a back-thrust or side-thrusting device to hold the force, are the only feasible system.

(vi)   *Snake movements.* These have not yet reached practicable commercial application but can have very interesting applications for moving through complicated systems of obstacles and grasping complicated shapes.

[†]K. J. Waldron and K. L. Kinjel, 'The relationship between actuator geometry and mechanical efficiency in robots', *4th CISM-IFToMM Symposium*, p. 366.

CHAPTER 7

Robots: the current state of theory and practice

7.1 THE STAGES OF ROBOT DEVELOPMENT

For the purposes of this chapter the successive generations of robots will be defined as follows.

First generation robot (or senseless robot). A machine which can be given a program by a human master to carry out any one of a variety of movements of a tool or of an object picked up and held in the robot's gripper (hand, or end effector). It can then continue to repeat exactly this programmed series of movements as many times as required; it has no sensory feedback and attempts to continue its movements regardless of changes in the external circumstances such-as the object to be manipulated being incorrect, wrongly orientated or even absent, or an obstacle being in its way.

First generation robots are either stationary or move on fixed rails as sensory adaptability is essential to free movement.

Second generations robots – simple sensory adaptability. These machines have one or more sensory devices giving yes–no binary information such as tactile contact, motion resistance, proximity sensors, light blockage sensors. They are programmed with simple algorithms to do the normal sequence if the information is 'yes' and to change to an alternative pre-programmed action if it says 'no'.

Such robots can be self-mobile on a working floor as they can use tactile or proximity sensors for fine positioning.

Third generation robots – complex information processing. These have the ability to process such information as a two-dimensional TV picture or multi-point scan and vary their movements accordingly. The earliest example was 'Shakey', the Robot built at Stanford Research Institute in 1968 (Raphael: *The Thinking Computer,* W. H. Freeman, 1976, p. 252). This had no hand or arm but could move freely about the laboratory floor and search for different-shaped objects (such as the largest cube or the smallest prism) and push them to a new instructed position.

Fourth generation robots. It is the author's belief that a true fourth generation robot cannot be developed since this requires the human faculties of free will, choice of alternatives accordingly to long-term motivation, and true originality. Such a robot would, like the most intelligent and wise human, be able to work out its own objectives as well as a plan of campaign to achieve them and modify this plan according to changes in the external situation.

7.2 INDUSTRIAL ROBOTS IN PRESENT USE

By 1981 — some 20 years after the first industrial robot, the Unimate, was on the market, Japan and the USA have both introduced quite large numbers of first generation robots (*CME,* May 1981, p. 21) with 6,000 and 3,500 respectively. The number in Japan is now rising by some thousands each year. Sweden and Germany have just over 1,000 each and the UK under 400.[†] The Socialist countries are believed to have about 500. The world total is probably some 15,000, which may be compared with the world total of manual workers of some 800 million. Thus the rate of growth in the 20 years since the first practicable robots were introduced has not yet been such as to make large changes in production except in certain special industries. Robots have been introduced mainly for the purely economic reasons that they are cheaper or more reliable than a human worker doing the same job. George Devol (*Industrial Robots,* 2nd edn, Vol. 1, p. 161) patented designs for an industrial robot in the 1950s to free workers from repetitive machine tending tasks, without suffering from the obsolescence of a special purpose device designed for a single task. It was to be capable of being reprogrammed for other manipulative tasks so that it could be transferred when the first task was no longer needed.

In 1958 Devol licensed the firm that became Unimation Inc. (President, J. F. Engelberger) to produce robots; three prototypes were tested in 1962 and 70 'Unimates' had been built and were working in factories by 1966. By 1974 600 Unimates were installed by Unimation and its Japanese licensee Kawasaki. Engelberger has been principally responsible for the overcoming of the many practical difficulties of producing robots that have a mean time between failures (MTBF) of 500 hours (J. F. Engelberger, *Robotics in Practice,* Kogan Page, 1980, p. 85) and a downtime of 2%.

Comparing the robot with 'hard automation' Engelberger claims (*Robotics in Practice,* p. 16) that in addition to the resistance to obsolescence of Devol's original aim, it has the advantages of being much more rapidly available ('off-the-shelf') and programmed for a new model or a different casting, and that the debugging which is always necessary on a complex engineering device has already been done on the robot.

[†]In his book *Robotics in Practice* Engelberger lists as responsible robot manufacturers three companies in Sweden, two each in Italy and England, one each in France and Germany, four in the USA and nine in Japan.

Apart from the saving of labour costs there are other potential advantages of robots, in comparison with human operatives. A robot in general uses more energy than a human for a given job, typically a robot might require about 5 kW even when the strength is no greater than that of a man. On the other hand it may save considerable energy by doing the job more accurately and not getting tired and producing rejects. Similarly it is not in general so fast-moving as the human hand can be, but again greater reliability may offset this disadvantage. However, by far the greatest advantage comes when it can do jobs which are unpleasant, uncomfortable, or dangerous for a human or when it is necessary to lift considerable loads. So far it has not anywhere near the same sophistication as the hand/eye co-ordination of a human and never will have without excessive cost because a human takes 10 years to learn to coordinate bodily movements with sense perceptions and especially the visual image, and several more years to learn a craft skill such as joinery or car repair. Moreover an intelligent human faced with a new job requiring repetition gradually works out the best tools and best methods for doing it by a combination of invention and empiricism.

Looking to the future it is probable that robot usage in manufacturing industry will level out in all developed countries at a figure of the order of one robot to every 1,000 human workers and the robots will probably be doing those jobs which are more or less repetitive but primarily dangerous and uncomfortable. This future development is discussed in more detail in Chapter 9.

Mechanical arms with hands without servo-control are often called robots. (W. R. Tanner, (*Industrial Robots*, Vol. 1, p. 6) calls them 'non-servo robots', The Japanese Robot Standardization Committee (*2nd CISM-IFToMM Symposium*, p. 495) calls them 'sequence robots'. These devices are also called 'pick and place', 'bang-bang', 'end-point' or 'limited sequence' (see Fig. 7.1). There are several such devices on the market, usually operated pneumatically, but they can also be electrical or hydraulic. They are cheap, can work with high speed and give good repeatability; they usually have a limited programme of movements set up by a rotating drum valve, or cams on a spindle rotated by an electric motor. Each axis of movement can have only two positions according to which end of a cylinder has the air pressure and the length of the movement is adjusted by end stops. Examples are the Prab, Mobot, Auto Place, Seiko. Additional intermediate stops can be provided on some axes which can be used on some steps of the program.

The control sequencer or stepping switch receives a drive signal from the limit switch of the previous movement, or from an external signal which tells it to end the movement earlier. It then goes on to the next operation. In the case of a uniselector switch such as was used in the table-clearing robot (Mark I, Fig. 7.2(a) built 1962; Mark II, Fig. 7.2(b) built 1962) or the rat in a maze[†] (Fig. 7.3)

[†]The 'rat' moved in sequence down each of the eight paths taking a choice of path at each junction but when a small object ('cheese') was clamped at the end of one path it locked onto this path and repeated it continually.

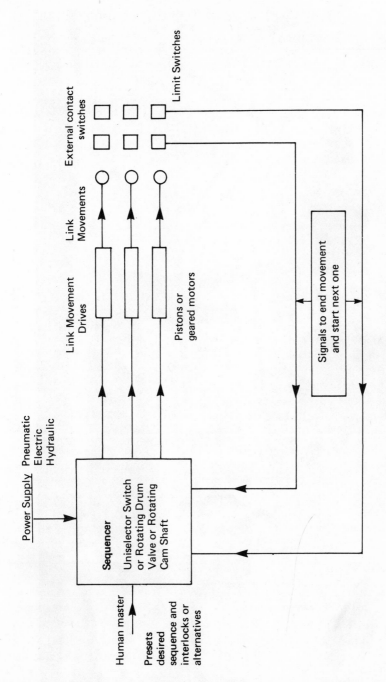

Fig. 7.1 – Diagram of limited sequence device.

Fig. 7.2(a) – Table-clearing robot, Mark I.

Fig. 7.2(b) − Table-clearing robot, Mark II.

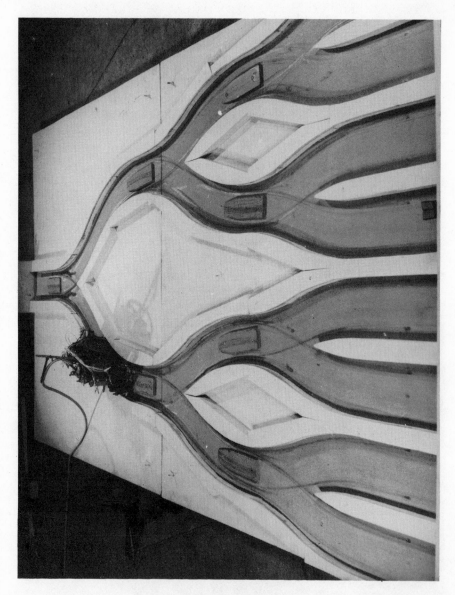

Fig. 7.3 – 'Rat-in-a-maze' robot.

a microswitch operated by an external contact can tell it to move to the same sequence or to a different one. When a limited sequence device is used to feed a machine the external signal will have to come from that machine.

The servo robot or high technology robot can be used for a much greater variety of industrial tasks which will be discussed in section 7.4. The arm and wrist usually has five or six degrees of freedom which we will write [5] or [6] : [5] when 1 wrist rotation is eliminated. The kinematics of the arm correspond to one of the four types shown in Fig. 7.4, the rectangular system or PPP involves linear motion along the three perpendicular co-ordinate axes; an example of this is the Italian Sigma (*Consideration of the Design of the Olivetti Sigma,* M. Salmon, *2nd CISM-IFToMM Symposium,* p. 93). The second type is the cylindrical or RPP, example the Versatran (*Industrial Robots,* 2nd edn. Vol. 1, p. 264). The third is the spherical RRP of which the larger Unimate is a famous example, which was first in the field. However, the most common is the jointed RRR which corresponds to the human arm although often the elbow bends downwards (as shown in Fig. 7.4, Type 4a) instead of upwards as in the human (Type 4b). Examples are the Cincinatti Milacron, the Puma, the Trallfa and the ASEA. The advantages of using linear motion, especially in the rectangular system (Fig. 7.4, type 1) are (1) it is not necessary to have a computer to interpret the *xyz* co-ordinates into the movements of the joints; (2) the arm inertia and leverage can be kept substantially constant so the same maximum weight can be handled in all positions. On the other hand the advantages of rotary joints are as follows.

(1) It is possible to have a much greater reach, for example a single prismatic joint can only extend to about 1.8 of the contracted length, whereas a hinged arm can bring the arm very close to the shoulder or take it away to the full length of the arm. It is also possible with the jointed system to bring the hand right in to the base of the robot.
(2) It can be designed to get round obstacles into awkward places.
(3) There are no exposed slideways which must be kept free from dirt and dust.
(4) All the rotary bearings are easy to seal.

This type of robot can have various interchangeable plug-in hands or it can replace the hand by a tool and it can be programmed to switch this tool on and off as part of the sequence.

Among the tools which have been fitted to robots are stud welding heads, inert gas arc welding torch, a heating torch to bake out foundry moulds instead of placing them in an oven, a ladle for molten metal, a spot welder, pneumatic nut runners, drills and impact wrenches, routers, sanders and grinders and paint spray guns, (Engelberger, *Robotics in Practice,* p. 55). By using a cradle for each tool and a standard correction system such as a bayonet or snap-on fitting, a robot can carry out a sequence of operations with different tools.

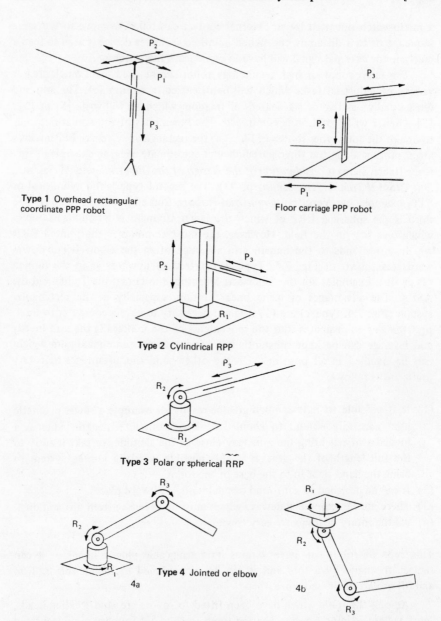

Fig. 7.4 – Principal types of robot arm or hand positioning systems in industrial use.

Three main systems have been used for controlling the movements of a robot (see Fig. 7.5).

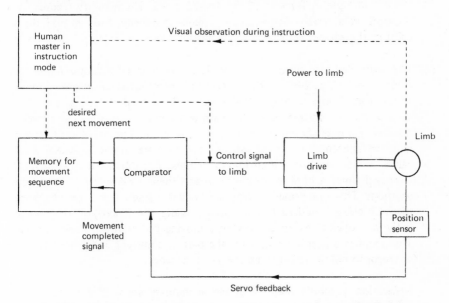

Fig. 7.5 – Block diagram of servo control system for one movement of generation 1 (senseless) robot shown in play-back mode (instruction mode shown dotted).

(1) *Point to point control.* This was the system used on the original Unimate, the path is not programmed, the instructor feeds in the co-ordinates of the end points of a movement and the machine then chooses its own path from one point to the next. If it is necessary to proceed via a path other than a straight line, to avoid potential obstacles for example, then suitable inter-mediate points must be instructed. If the point to point control is digital, i.e. each point is defined by a set of numbers corresponding to the location of all the drive positions to reach that point, then the memory can carry several hundred point positions and the robot can not only do any operation that a limited sequence device can do but also much more sophisticated tasks such as putting objects into the positions of a pallet, stacking or spot welding.

(2) *Continuous path.* When it is necessary to follow a particular path from one point to the next, as for example in paint spraying, (where the sprayer must be held at a standard distance from the object being sprayed) or arc welding or milling round an edge, the continuous path control system is used. The operator leads the hand manually through the required path and the computer stores the instantaneous position of all the axes corresponding to a closely spaced succession of points on the path. The continuous path robot can be

taught to have the required speed of movement as well as path. The teaching operator moves the hand in real time at a speed which is the same or a chosen multiple or fraction of the desired speed. The memory records the positions at a certain frequency, e.g. mains frequency, and can play them back at the same frequency or at a multiple corresponding to the present figure.

(3) *The controlled path system.* An example is the Cincinatti Milacron (*Industrial Robots,* Vol. 1, p. 296). In this system the operator has co-ordinate control of the Tool Centre Point (TCP) and can set the desired velocity and acceleration along the desired path. The path is normally a straight line between a series of points chosen so that if a curve is required points closer together must be chosen. The co-ordinates of the TCP are stored in spatial XYZ values so the control system knows in these coordinates where the TCP is in real time. A built-in program converts these XYZ values to arm joint positions. This arrangement is especially valuable where it is desired to move the TCP along a desired line in space (usually the Y axis) so that it can carry out operations on objects on a moving belt without itself moving. This requires a position sensor for the part of conveyor and a limit switch or sensor to tell the robot when the part is in range.

Generation 1 robots that are in use in industry are mostly hydraulic or electrical. Pneumatic drives are cheap and reliable but are not good for servo control owing to the compressibility of the air. Electrical systems used are servo motors, stepping motors, pulse motors and solenoids: these are used in about 20% of robots; the motors are high-speed, low torque and require gearing or screws. Hydraulic piston drives are simple and reliable, give no fire hazard from sparks (although oil may have to be replaced by non-flammable working fluid in situations where a small leak could be hazardous) and can use an accumulator to give three or four times as much power for a rapid movement as the pump power.

In general smaller robots are most economically electrically driven while the large ones that work in more dirty and rough surroundings are usually hydraulic. Servo-valves are nearly as costly when small as when large, unlike electric servo amplifiers.

When choosing a generation 1 robot for industrial use the main factors to be considered are as follows.

(1) The weight of objects to be handled — robots maxima range from 4.5 to 800 kg.
(2) Working volume — the reach ranges from 0.5 to 3 m and the volume from 0.25 to 28 m^3.
(3) The type and complexity of task. Number of operations and special requirements such as continuous path or controlled path. Accuracy of positioning:

for example an arc welding robot needs great accuracy but not high speed, whereas paint spraying needs high speed but low accuracy.

(4) The conditions of operation, e.g. dust, heat, fire hazard.

7.3 EXISTING SECOND GENERATION INDUSTRIAL ROBOTS WITH LIMITED SENSORY ADAPTABILITY

The first generation robots described above have a defined positional accuracy but only repeat within this accuracy the exact movements they have been programmed to to. To grip an object requires more precise positional accuracy than the 0.7 mm of a conventional movement and if the object itself has slight variation in size further adaptability is required. A human picking up an object uses the eye to estimate approximately the position and size, touch to give exactly the position, and muscular force feedback to give the exact grip required to pick up the object. We can pick an object up by grasping parallel sides and adjusting the grip according to the friction. Generation 2 robots are provided with simple touch force or proximity sensors (Fig. 7.6) so that they can modify their actions or undertake a branching program when required by an observed change in external circumstances. An example of a branching program is when the tool can be sensed as becoming blunted and the robot stops the use of the tool and enters a branch program in which it changes the tool. The use of the

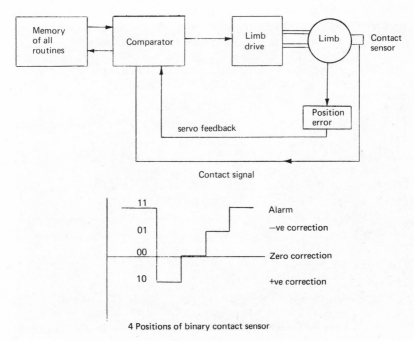

Fig. 7.6 − Control system for generation 2 robot.

microprocessor has made it possible for a robot to have its own individual program responding to its sense observations, either by altering its movements or by going into a branching program. There are a number of different commercial makes of robots with this possibility (e.g. Versatran, ASEA, Unimate and Cincinatti Milacron.)

An early example of a sensory adaptive robot was Thring's table-clearer (Fig. 7.2) which determined the position of an object on a table as the robot traversed along the table by the interruption of a light beam observed by a single photo-cell; then the arm came out until it detected the object by a light touch on the gripper which operated a microswitch and the hand then closed until a firmer pressure on another microswitch indicated the grip of the object, which was then cleared away on to a revolving tray. Another early experiment was a continuous arc seam welding device which sensed the position and width of the gap between two plates by using the pressure of the welding gas jet reflected off the sides of the gap and sensed on each side of the tool, developed at QMC for the Welding Institute.

Robots with adaptability of action or branching choice of program based on touch sensitivity are in regular use mainly with simple contact sensors to adjust the position of a robot-held tool over small variations, or to follow a curve. Arc welding and deburring are cases where the technique is valuable (*'Giving robots the power to cope'*, H. Colleen, *Robotics Today,* Spring 1980).

Fig. 7.7 – Sheep shearer robot.

Force sensors are still in the experimental stage and will be described in section 7.5 as they are being used for advanced assembly experiments.

J. Trevelyan ('Automated sheep shearing', *Seminar Robots in Manufacturing Industry*, 1981) has developed an experimental shearing rig (Fig. 7.7) at the University of Western Australia (Perth) for the Australian Wool Corporation which is one of the most sophisticated sensor-adapted robot manipulations in the world. The dirty, hard physical work of sheep shearing is a major bottleneck in the Australian wool industry and it takes a man about 10 years to become a skilled shearer. The work has shown that automated mechanical shearing is technically feasible but it is unlikely ever to be economic on flocks of less than 4,000.

Figure 7.8 shows the way in which the cutter is held to enable it to be orientated correctly on the sheep's back. The three axes of rotation intersect

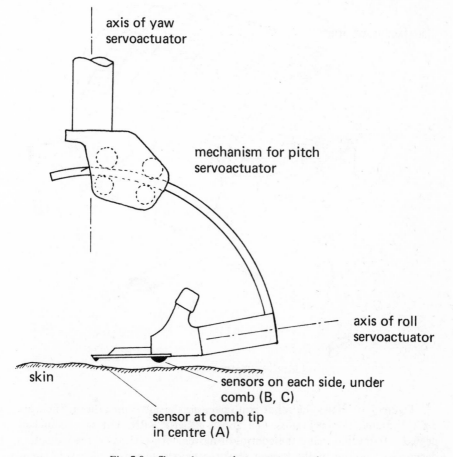

Fig. 7.8 – Sheep shearer robot: cutter control.

so that changes in orientation do not affect the position of the cutter — a good example of the use of mechanical design to simplify the control system. Three capacitance sensors are mounted on the cutting comb and provide fine positioning and orientation of the cutter a short distance from the sheep's skin. The rough positioning and orientation are done by a 'software sheep' that is a relatively simple model of a typical sheep (see Fig. 7.9) with two surface coordinates similar to latitude and longitude on the spherical earth. The sensors provide observations on the difference between the actual sheep and the software sheep which are used to modify subsequent cutting blows.

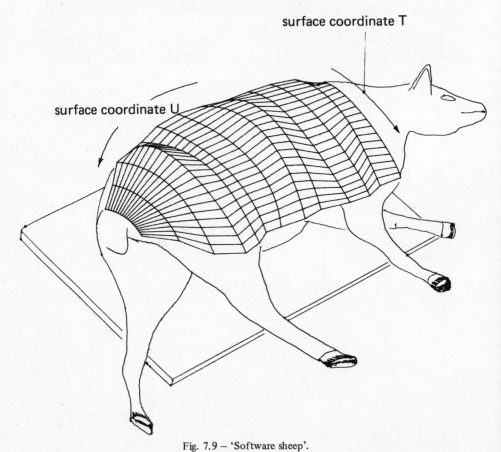

Fig. 7.9 – 'Software sheep'.

Figure 7.10 shows the robot arm operating the cutter on a sheep. There are eight hydraulic servoactuators, the arm is PRRRP RRR. The two redundant degrees of freedom make the control of movement easier as do the redundant degrees of freedom of the human arm; the twist between shoulder and elbow

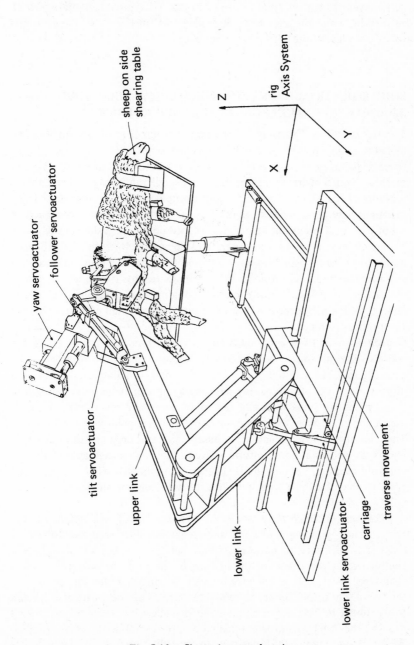

Fig. 7.10 – Sheep shearer robot rig.

and the movement of the shoulder pivot point which enable us to reach the same hand position by different routes. In the shearer case for example the second P movement (called 'follower servo' in Fig. 7.10) provides radial approach to the surface, while the associated R (called 'tilt' in Fig. 7.10) gives a direct control of the angle of the cutter to the surface.

7.4 SOME INDUSTRIAL TASKS FOR WHICH COMMERCIAL FIRST AND SECOND GENERATION ROBOTS ARE USED AT PRESENT

(1) *Material handling.* This is by far the largest group of robot applications and includes taking blanks or partially finished objects or completed objects from or to stores or conveyors or containers and to or from various types of machines which operate on these blanks. A single robot may load and unload a group of machines or may be fully occupied with one machine. These machines may be in series or parallel as far as the blanks are concerned. A robot can load bricks onto a pallet or shaped objects into slots in the pallet. These are extremely routine operations in which the robot is primarily useful in saving humans from boredom and fatigue which cause unreliability and sometimes danger. Among the machines fed by robots are stampers, die casting, investment casting, plastic moulding, metal cutting (e.g. lathes, millers, drills). First generation robots have been mounted on tracks to serve 10 or more machines.

(2) *Surface coating.* Paint spraying is an extremely unpleasant task. There is a real fire hazard and in some cases some paint sprays are believed to be carcinogenic. The painting room must be small to maintain dust-free conditions. The use of robots for paint spraying can reduce the ventilation requirement by some 60% and the robot can make much more accurate use of the paint once it is programmed not to waste paint beyond the edges. This can give a 40% reduction of paint usage and only one-third as many rejects, as well as a more uniform coating. The robots used are mostly not sensorily adapted, i.e. they are first generation robots, usually of the jointed type, so that they have to be fed with the work piece in an exactly identical position each time.

(3) *Applications in which the robot holds a rotary tool.* These include drilling and routing, grinding, polishing and deburring. For accurate drilling of holes it is necessary to have a template with bushed holes in it and some compliance in the robot for the holding of the tool, the hole must have a chamfered entry sufficiently wide to guide the drill in within the accuracy of the robot's position. Similarly template surfaces are necessary if a plate is to be routed to have very accurately shaped edges but a robot is often used to move the tool by its programmed path only, when less accuracy is needed. Where a large number of holes have to be drilled on a single component a

robot is often more economical than either a human or a special purpose multi-hole drilling device, especially if the component is not flat so that the holes must be drilled in different directions. This was found to be the case in the aircraft industry, ('Robots drill and profile', K. Gettelman, *Industrial Robots,* 2nd. edn. Vol. 2, p. 357). Aluminium alloy fuselage components are slightly different, each can be drilled and profiled to its own specification from the robot program which is selected by an optical character readout.

(4) *Welding.* The use of robots in spot welding of car bodies is one of the best known economic applications because each body requires a large number of spot welds, accurately and reliably placed and it is a straightforward matter to use a point to point Generation 1 robot without senses programmed to put them all in the same place on each body. In spot welding the change in current in the locating motors of the robot arm, when the spot welder contacts the metal, is used to provide the signal for switching on the welder. Continuous path robots have also been used for automatic arc welding and in some applications, (*Industrial Robots,* 2nd edn. Vol. 2, p. 305) the robot has a position sensor just ahead of the weld gun and the robot can have a weave operation to fill a wide joint.

(5) *Inspection – quality control.* Robots have been used to check the dimensions of car bodies by pushing probes at 100–200 points on the body ('Robots combine speed and accuracy in dimensional checks of automotive bodies', G. C. Macri and C. S. Calengor, *Industrial Robots,* 2nd edn. Vol. 2, p. 41). The data for 3,000 checked bodies are stored on a computer with graphics capability. Auto-place have produced a generation 3 robot system using four solid state image cameras with 60,000 picture elements to tell the robot whether to accept or reject engine valve covers ('Smart robot with visual feedback boosts production 400%', *Production,* June 1979). The covers are also tested for leakage by charging with air at 5 psi.

(6) *Assembly.* Assembly is a highly sophisticated human task and first and second generation robots can be used for this purpose only with the help of devices such as bowl feeders and pallets, and where the joining processes are simpler than the use of screws or rivets. Special programs aimed at advanced assembly will be described in the next section.

The Sigma robot (M. Salmon, 'Considerations of the design of the Olivetti Sigma', *2nd CISM-IFToMM Symposium,* p. 93) uses the PPP movement to locate a gripper which hold a drill or a pin for the assembly of household appliances. Bushes with conical entry in a mask are used for fine location of drilling points, and springs in the wrist allow $X-Y$ errors. A third spring and transducer observers the Z axis (vertical) force. Only one component of force is read at a time, that in the direction of the assembly operation to be performed, e.g. inserting a pin along the Z or X direction. The force information is used to modify the subsequent movement of the gripper.

The PUMA robot has been specially designed by Unimation in conjunction with General Motors for the latter's Programmable Universal Machine for Assembly ('Assembly robot update', *Industrial Robots,* 2nd edn. Vol. 2, p. 462). The robot arm is RRR and the wrist RR so it is only 5 axis and can handle parts up to 2.3 kg with a positional repeat accuracy of ± 0.05 mm. Its use for a number of simple assembly operations is being studied. The difficult assembly problems are still parts feeding and orienting ('Puma robots find cost-effective jobs in batch manufacturing', *Industrial Robots,* 2nd edn. Vol. 2, p. 469), and the initial applications will be Generation 1 using dead reckoning and accurate fixturing. Generation 2 additions of tactile and visual feedback, acoustics, lasers and proximity sensors are being studied.

7.5 EXPERIMENTS ON ADVANCED ASSEMBLY

Several groups have been studying the possibility of using generation 2 or generation 3 robots for sophisticated assembly of a repetitive character. A human can be trained quite rapidly for such tasks as assembling small electric motors because of their trained hand—eye coordination and even a blind person can learn to do it with a highly developed touch sense, but the problems for a robot are:

(1) to find the chassis and locate it with the correct orientation in a vice or special holding fixture;
(2) to find and orientate each component;
(3) to check components for common errors;
(4) to fix the component in place by means of pin, screw (self-tapping or with threaded hole or nut), push fit (e.g. commutator on shaft), solder, weld, bent lug, jubilee clip.

An extreme example such as assembling the front plate of a clock mechanism with all the drive shafts in the right holes, taxes the skill of a highly trained mechanic and has already been designed out of mass-produced articles; but inserting a fine screw so that the thread engages correctly requires a very sophisticated combination of sight and fine force sensing.

In the USA the National Science Foundation has been supporting research work for some years at Stanford Research Institute and Charles Stark Draper Laboratories. J. L. Nevins, D. E. Whitney and colleagues at the Charles Stark Draper Laboratory have studied in detail the insertion of a peg in a hole ('The force vector assembler concept', J. L. Nevins and D. E. Whitney, *CISM-IFToMM Symposium,* Vol. 2, p. 273; 'Results of programmable assembly machine configuration', A. S. Kondoleon, *Industrial Robots,* Vol. 2, p. 542) (see Fig. 7.11).

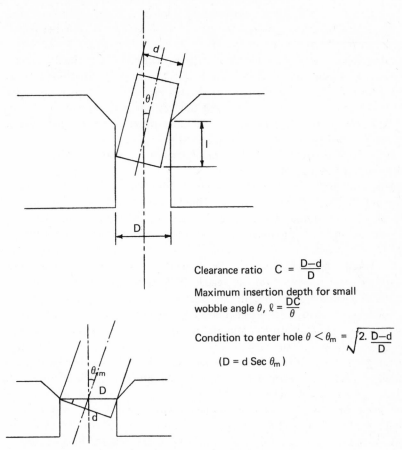

Clearance ratio $\quad C = \dfrac{D-d}{D}$

Maximum insertion depth for small
wobble angle θ, $\ell = \dfrac{DC}{\theta}$

Condition to enter hole $\theta < \theta_m = \sqrt{2.\dfrac{D-d}{D}}$

$(D = d \, Sec \, \theta_m)$

Fig. 7.11 – Geometry of peg in hole.

This is perhaps the simplest assembly process as the insertion of a screw into a tapped hole requires greater accuracy of alignment if the thread is to mesh. One way of doing this process is to make sure that the robot hand holds the peg accurately vertically and that the hole is accurately vertical also $(\theta = 0)$. This reduction of the degrees of freedom of the system, however, limits the jobs. The human hand is capable of inserting a screw into a tapped hole at any angle to the vertical, although the problem is much more difficult if one cannot use the eyes as well as the sense of feel to ensure alignment and location. The problem with which the team is concerned, therefore, of the robot hand having to obtain both sufficiently accurate alignment and location is one of very considerable interest. Whitney assumes that there is a chamfer at the top of the hole or on the end of the peg which is slightly greater than the maximum displacement error so that the end of the peg will reach the chamfer and slide down into the hole.

He also assumes that the peg (diameter d) is a loose fit in the hole (diameter D), so there is a substantial clearance $(D - d)$. If there were no friction between the contact at the end of the peg and the chamfered hole the peg would slide readily down the chamfer provided the robot wrist has compliance in the directions of the plane of the surface while holding the pin parallel to its first position. The contact force would be at right angles to the chamfer and thus would have a substantial component moving the peg sideways in the required direction to reach the hole. Friction, however, reduces the lateral force and increases the axial force making it more difficult to sense the required direction of movement for location. With steel the coefficient of friction is about 0.2 whereas it is 1 with aluminium because it is softer.

When the peg is inserted with a slight obliquity the contact force produces a moment tending to straighten the peg. Thus in general a wrist which is compliant, both in regard to rotation and to parallel lateral movement can be used to get the peg into the hole provided the errors are not greater than values fixed by the size of the chamfer and the degree of clearance. However, the rotation to obtain correct alignment must not be about the centre of the wrist but about the end point of the centre of the peg. P. C. Watson describes a remote centre compliance, RCC, which gives this type of movement by a combination of a rotational part which is trapezoidal and a parallelogram translational as shown in Fig. 7.12(a) (P. C. Watson, *Industrial Robots*, Vol. 1, p. 414). A peg can be inserted in a hole without chamfer by tilting it and moving it along until the lower edge engages with the hole. The remote axis admittance or remote centre compliance does not measure the errors it encounters and so it is not possible to have a force sensor feedback loop to move the wrist so as to reduce the error. Such force information can also be used to teach the robot to improve its technique. The Draper Laboratory carried out a study[†] on an automobile alternator assembly with a force sensor in the pedestal which displayed the three torques and three linear forces to an operator and it was found that this information was of real value in the robot assembly teaching process. By measuring the two linear displacements of the plate A (Fig. 7.12a) rigidly connected to the pin and the two rotations about axes in its plane they could obtain measurements proportional to the forces due to the displacement and rotational errors. It should be noted, however, that the RCC is light and has rapid response so that tight insertions can be made in 0.2 secs, whereas an active servo feedback system has to move the robot arm and requires 3–5 secs.

The other main laboratory working on programmable assembly robot systems in the USA is the Stanford Research Institute. They are working on a programmable generation 3 system of presenting parts to the assembly with the correct orientation using vision to determine the stable state and orientation of the part

[†]D. S. Seltzer, 'Use of sensory information for improved robot learning', *Industrial Robots*, 2nd edn. Vol. 1, p. 424.

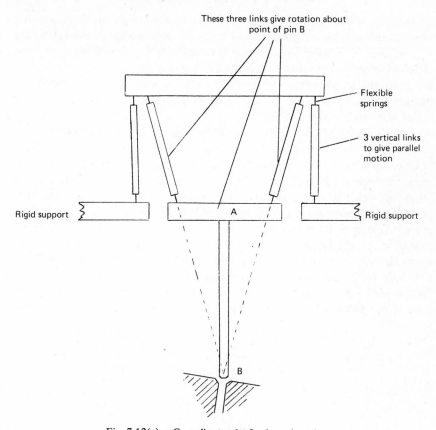

Fig. 7.12(a) – Compliant wrist for inserting pins.

(J. W. Hill and A. J. Sword, 'SRI Reports on its programmable parts presenter', *Industrial Robots,* 2nd edn. Vol. 2, p. 470). The part is fed individually by a belt or vibratory feeder onto a translucent shuttle tray at position 1 of its three positions. The shuttle carries it to the central position (2) where its contour is viewed with back lighting by a TV camera. The image is processed to determine which of the possible states the part is lying in and its orientation (length and orientation of major axis and maximum radius). If the image does not match any of the shapes previously fed into the computer, the shuttle pushes the part to position 3 which is a reject chute. If it does match one shape the turntable on the shuttle is rotated to a predetermined position such that when the part is tumbled over a step of a predetermined height it will stand with the correct orientation on the correct face. An elevator surface under position 1 is then lowered to give this step height and the shuttle pushes the part onto this surface. The elevator is raised and the part pushed back onto the shuttle which returns to position 2 for centring and reorientation. Finally the part is picked up by a limited sequence device and placed on the pallet which is on an $X-Y$ table and moves to the next vacant space after each placing. This process requires a number of servo-controlled mechanical movements and it will clearly take a great deal of development work to do this task as quickly as a human.

J. W. Hill of SRI (Stanford Research Institute) ('Force control assembler', *Industrial Robots,* 2nd edn. Vol. 2, p. 474) describes a three-axis force controlled wrist with movements of 25 mm and locking cylinder to hold the x and y axes at the centre of their range when desired. This wrist can be mounted on a conventional robot arm and each movement is servo back-drivable with a force capacity of 40 kgf and a sensitivity of 0.2 kgf. Used in conjunction with an impact wrench it can pick up a bolt from a feeder with a conical guide which drives the socket to the correct position for pick-up and insert it into the head of an air compressor moving down an assembly belt. The robot arm gives approximate following of the movement and the wrist corrects it. There is no rotational control so that the bolt and compressor must be accurately vertical. Figure 7.12(b) shows the strain gauge wrist sensor built at SRI which enables the three force and three torque components to be measured with 7-bit accuracy (1 in 128). This consists of three aluminium rings (machined from 75 mm tube) the top one being joined to the arm and the bottom one to the hand. The top one is joined

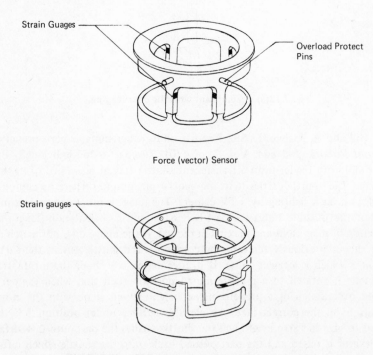

Fig. 7.12(b) – SRI strain gauge wrist sensor (six axis).

to the middle one by four vertical narrow beams with an elastic neck at one end, and the middle one to the bottom by four similar horizontal beams. Strain gauges one each side of the upper beams give the torque signals and on the lower beams the linear force signals.

The Westinghouse R and D Center has been working closely with the Draper laboratories and SRI to study the commercial practicability of various advanced assembly robot operations ('Cost-effective programmable assembly systems, R. G. Abraham and J. F. Beres, *Industrial Robots,* 2nd edn. Vol. 2, p. 429). They conclude that complex assembly tasks are more appropriate for humans and are not even technically feasible at present, and that the number of programmable automatic assembly applications will remain small until lower cost robots work faster than humans, and simple low cost parts presentation methods become commercially available. A mix of man, programmable and fixed automation equipment can improve the system cost-effectiveness.

Heginbotham and his co-workers at Nottingham have also developed a range of devices for programmable assembly machines. In 1970[†] he described the Minitran, which is a very cheap limited sequence device with a single arm. The idea was to have a number of these assembled as a group; each would handle one component and place it on the assembly and then pass the assembly on to the next. Each unit was sufficiently versatile to be changed over for a different component, as necessary. Thus a complete assembly system could be made of a number of these without having to have a rotating table or moving belt. The group would be operated by a central electronic system which would synchronise and interlock them.

In 1974 he and co-authors described an experimental programmable assembly machine[‡] which could have versatility for small batch production, the components to be assembled were placed in magazine tracks, sloping at 45° along the back of the machine and screws for joining them were in a magazine from which they could be picked up by a power screw driver. The machine had three linear movements but no rotations. Movement of the table gave the Y axis displacement and the arm carrying the placement device (gripper) gave the X axis movement and the vertical displacement Z. The arm also carried the power screwdriver in parallel with the gripper.

In 1978 Heginbotham and co-authors[§] described a fully operational versatile various mission assembly machine. This could assemble a telephone relay by fitting on to the relay body any distribution of up to 20 springs picked out from

[†]W. B. Heginbotham, 'Automatic assembly tomorrow', *Production Engineer,* July 1970.

[‡]W. B. Heginbotham, D. W. Gatehouse and A. Pugh, 'Programmable assembly machines – have they a future?', *2nd Conf. on Industrail Robot Technology,* Paper A1.

[§]'A versatile variable mission assembly machine', W. Heginbotham, A. Pugh, D. W. Gatehouse and D. Law, *3rd Conference on Industrial Robot Technology and 6th International Symposium on Industrial Robots,* 1978, A5, p. 54.

a magazine with 36 stations for different springs. The springs are brought in magazines suspended from rods, directly from the press tool that made them. The machine was instructed which sequence of springs to use by means of a brass tape with slots giving a code with 6 binary channels: infra-red light was shone through the slots onto a reader. The machine had a vacuum pick-off which could pick up the required spring and rotate it through 90°; then the spring was carried by the gripper on the pick and place carriage to the platen where they were assembled. The platen had a main movement and a fine top movement for the position of successive springs on it. The average spring placing time was 4.5 secs. The fast forward or backward movements of platen carriage and pick and place carriage were obtained by having, for each movement, a motor running in each direction, either of which could be coupled with clutches directly to the movement; a brake was applied to give fast stopping when the system was de-clutched. The lead-screw for the small movement of the assembly after each spring was operated by a Geneva mechanism, one turn covering the required movement.

While this work is extremely ingenious it would seem that humans will always be capable, fast and efficient for tasks of complexity, sophistication and short runs. There will, however, be a certain number of repetitive assembly tasks which can be redesigned to make them suitable for robot assembly. Highly specialised variable assembly devices of the limited sequence type as described by Heginbotham clearly have a future for this specialised work as the human brain is not well adapted to the rapid selection of a different sequence every time.

7.6 THIRD GENERATION ROBOTS WITH SHAPE OR PATTERN RECOGNITION

A great deal of research is going on in this area but it has not yet reached the stage of reliable commercial application and it is doubtful whether it will ever do so because of the fundamental law of engineering artefacts that *as the system becomes more complex the cost to make it function reliably with a long mean time between failures (MTBF) rises exponentially.*

A block diagram of a complete third generation robot is shown in Fig. 7.13. This is still a robot because all its actions and decisions are based on the prior instructions of a human master but it can take complex decisions based on the combination of sense observations of three-dimensional reality by touch, force and two dimensional pictures.

The nearest robot to a complete third generation robot is probably the Waseda robot, Fig. 7.14, which has two TV camera eyes in its body and can walk on two legs balancing by moving its weight (I. Kato *et al*, 'Information power machine with senses and limbs', *1st CISM-IFToMM Symposium*, p. 11).

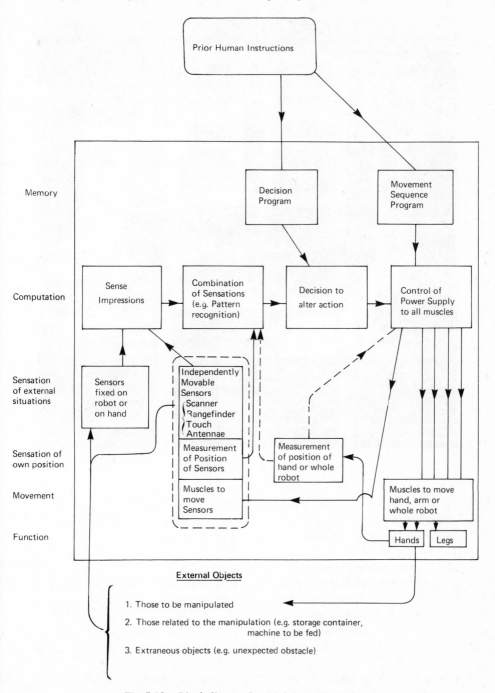

Fig. 7.13 – Block diagram for third generation robot.

Fig. 7.14 – Waseda robot.

Some of the tasks that a human can do quickly and reliably because of the sophisticated processing of visual information and the co-ordination of complex hand movements to visual and tactile senses are:

(1) to pick a randomly oriented object out of a bin;
(2) to choose and locate a rarely used particular tool or component from a workshop or obtain one from a storage system, e.g. drawers of nuts and bolts, wood screws, rivets etc.;
(3) to find which component in a machine has failed and replace it or repair it;
(4) quality control of three-dimensional objects which may have errors of an unexpected type;
(5) to develop an improved way of doing a job, or design an improved tool or jig.

Many programs of research have been carried out on the possibility of enabling robot sensing (third generation) of two-dimensional and even three-dimensional shapes, contours, silhouettes or profiles. These are based on one of the following:

(1) tactile or proximity sensing with two-dimensional sensor array;
(2) TV camera scans;
(3) two-dimensional light sensor array;
(4) one-dimensional light sensor array combined with physical movement of the object.

7.6.1 Tactile or proximity sensing arrays

Bejczy ('Effect of hand-based sensors on manipulator control performance', *Mech. and Mach. Theory,* 1977, **12**, p. 547), describes a 4 × 8 matrix of proportional tactile sensors made from two nets of electrodes separated by conductive rubber which can be fitted to each jaw of a parallel jaw hand to give limited shape information. Garrison and Wang built a gripper with 100 pneumatic snap-action touch sensors located on a 25 mm square grid ('Pneumatic touch sensor', *IBM Technical Disclosure Bulletin,* **16**, 6, Nov. 1973).

Heginbotham *et al.* (N. Sato, W. B. Heginbotham and A. Pugh, *Proc. 7th Int. Symp. on Industrial Robots,* Tokyo, Oct. 1977, 'A Method for three-dimensional part identification by tactile transducer') have described an 8 × 8 matrix of probes which can be used to give the three-dimensional contours of an object by gradually lowering the probe array onto the object and detecting the position at which each probe is first displaced by contact with the object. The probes were fine diameter steel pins moving in perspex tubes against soft phosphor-bronze springs and the first movement of 1 mm of any one probe was detected by having eight energising coils and eight sensing coils connected at right angles on the array. The use of long perspex tubes reduced the risk of a pin jamming when it contacted a steep, sloping surface on the object.

J. W. Hill and A. J. Sword of SRI ('Manipulation based on sensor directed control: An integrated end effector and touch sensing system', *Proc. 17th Annual Human Factor Safety Convention,* October 1973) made a parallel jaw hand with a matrix of 6 × 3 force-sensing buttons on the inside of each jaw.

S. Takeda ('Study of artificial tactile sensors for shape recognition: algorithm for tactile data input', *Proc. 4th Int. Symp. on Industrial Robots,* Tokyo, Nov. 1974, p. 199) built a device with two parallel fingers each with an 8 × 10 needle array. Contact of each needle with the object could be observed as the measured closing of the hands proceeded.

7.6.2 TV camera scans

The earliest sophisticated robot which could carry out actions based on computer processing of its TV camera scan was Shakey built in 1968 at Stanford Research

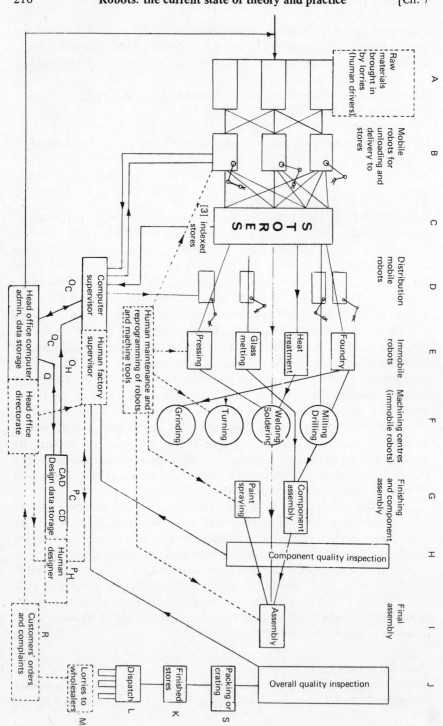

Fig. 7.15 – Fully automated factory for mass-production of a variety of similar equipment (e.g. cars, refrigerators, radios, TVs).

Institute (Fig. 7.15). This self-mobile robot had no arms but could recognise cubes, prisms and spheres of various sizes and push them around, avoiding obstacles, to instructed positions.

Many studies have been made since then to try to give vision to industrial robots so they do not have to work blindly. The Heginbotham SIRCH robot used a TV camera with 128 x 128 points taken from the standard 625 two-dimensional line scan to observe the two-dimensional silhouette of a black object against a white background ('The Nottingham 'Sirch' assembly robot', W. B. Heginbotham, D. W. Gatehouse, A. Pugh, P. W. Kitchen and C. J. Page, *1st Conf. on Industrial Robot Technology,* 1973, paper R9). The camera sees the whole work table vertically from above and when it observes an object is on the table, the camera (which is on the same turret head that carries the gripper) is moved to be over the centroid of the object. It then comes down into finer focus and so the object now fills the view.

A silhouette examination is also provided by the modular system, MODSYS developed in Karlsruhe (*3rd CISM-IFToMM Symposium,* p. 399, 'A modular system for digital imaging sensors for industrial vision', J. P. Foith, H. Geisselmann, U. Lubbert and H. Ringhauser). A black silhouette of the object which is three-dimensional and may be in several discrete positions, as well as any orientation on the plane, is used for determining the location and orientation of the parts for sorting them and quality inspection. A certain number of TV lines are selected and scanned for black and white points to determine the edges of the silhouette. The silhouette can be analysed for area, shape, moments of inertia, bounding rectangle, holes or indentations. During the using phase the same features as in the learning phase have to be extracted and compared with the stored features. The system has been used to calculate the area of biscuits to find it they are complete and to compute the volume of axially symmetric objects such as drops of molten glass using 1 m second exposure and summing the squares of the lengths of the linear intersection at four points along the drop.

Mitsubishi ('Where goes industrial robot technology', W. B. Heginbotham, *Production Engineer,* March 1976, p. 125) have used a TV camera in the hand of a robot to scan an object by moving and rotating the hand but the recognition period with the slow mechanical scan was 20 secs.

7.6.3 Two-dimensional light sensing arrays

An example of a recent experimental study using the computer vision for potential industrial robot applications is the use of an SRI international vision module by an aircraft firm for automated fastening of aircraft structures ('Robotic drilling and riveting using computer vision', R. C. Movich, *Industrial Robots,* 2nd edn. Vol. 2, p. 362). The SRI vision module (G. J. Agin and R. O. Duda, 'SRI vision research for advanced industrial automation', *Proc. 2nd USA–Japan Computer Conference,* 1975, p. 113) uses a 100 x 100 electro-optical sensor array to determine the identity, position and orientation of each one among a range of objects so that a robot arm can assemble them. It can also be used for simple inspection.

The NBS vision system ('Experiments in part acquisition using robot vision', R. N. Nagel, G. J. Vandenburg, J. S. Albus and E. Lowenfeld, *Industrial Robots*, 2nd edn. Vol. 1, p. 369) uses a solid state video camera with 128 × 128 sensors, and a stroboscopic light, both mounted at a fixed angle on the robot's wrist. The columns of the image sensor are perpendicular to the scanning flash plane, so each column of the image produces two numbers, the first the count from the bottom of the column to the flash line and the second the width of the image on the flash line. By angling the camera and light source at a fixed angle the distance of the object from the camera will change the relation between light plane angle and camera observed angle so that the distance can be measured while the shape can be computed from the 128 pairs of line width measurements. The system has been used to acquire and sort a pile of randomly oriented rectangular and cylindrical parts. The vision system operates in three stages.

(1) A faraway view is used to locate and move the robot up to the pile.
(2) A close view of the pile enables the vision system to isolate a single part, determine its orientation and identify the side of the part facing the robot. The control computer contains the data on all the part types and dimensions and uses this to interpret the scan data.
(3) The vision system selects closest side and the robot wrist moves till the grip fingers face this side head-on. It measures the side length and moves up for an overhead view to have a second side measurement. It can then pick up the block or cylinder and place it in a fixed orientation in a vice.

The work is fascinating and shows great ingenuity but a comparison with the skilled human mechanic suggests that the third generation robot will always be restricted to a proportion of tasks relatively simple and repetitive in numbers too boring for a human and insufficient for a fully automated device capable of this particular task alone.

7.6.4 One-dimensional light sensor arrays combined with perpendicular movement of the object

The Nottingham University group developed a method of using an untooled belt feeder in place of a bowl feeder specially tooled for the particular part. They used a line scan camera to observe the profile of the part as it moved. ('Versatile parts feeding package incorporating sensory feedback', A. Pugh, W. B. Heginbotham and K. Waddon, *Proc. 8th Int. Symposium on Industrial Robots*, Stuttgart, p. 206). As the profile of each part was observed it was compared with a stored 'mask' of the profile of the component with the correct orientation. Components with incorrect orientation can be recycled or orientated suitably programmed by a robot. Two line-cameras at right angles could be used. The camera has a linear array of 128 photodiodes. Components were characterised by grid pattern, area and perimeter, convex deficiency and external discontinuity.

The Consight system (M. R. Ward, L. Rossol, S. W. Holland and R. Dewar, 'Consight: a practical vision based robot guidance system', *9th Int. Symp. on Industrial Robots,* March 1979 and *Industrial Robots,* 2nd edn. Vol. 1, p. 337), has been developed by General Motors to pick up parts randomly placed on a moving conveyor belt based on SRI work. A 256-point line camera observes the orientation and position of a part as it moves along on the belt which is encoded for position and speed. The robot arm tracks the moving part, picks it up and transfers it to a predetermined location with a predetermined orientation. The parts must not be touching but two parts can pass together. No special light table or part colouring is necessary; two light lines from sources at an angle on either side of the camera produce a single line on the belt but both are moved out of the camera vision line when a thick object comes under the camera. Two are used so that the shadowing effect at the beginning or end of the object does not appear as object. From the camera information the x, y co-ordinates of the centre of silhouette area and the orientation and sense of the axis of the least moment of inertia are calculated. Parts which have multiple stable positions require multiple models.

7.7 ROBOT-OPERATED FACTORIES

The fields at the present time where robots are mainly replacing humans in factories mass-producing complex engineering products are:

(i)   Where the task can actually be done better by a robot with its greater strength and controllability; this includes welding processes on car bodies, paint spraying and handling rotary devices for routing and drilling.

(ii) Cases where the processes could be automated by special tools if much larger quantities of identical products were required but where robots offer the advantage of enabling one to vary the product quite frequently, that is with batches of the order of a few tens to a few thousands. These numbers are already unduly repetitive and boring for a human operator but not sufficiently so to justify the very large cost of special automation. This group may be called automated small batch production.

We can trace a whole sequence of steps leading towards the totally unmanned factory. These are as follows:

(i)   Soon after World War II the first numerical control systems were developed in the United States and in the United Kingdom. These were mainly either lathes or one- or three-axis milling machines. Some of them were continuous path, that is to say three movements at right angles could be controlled simultaneously to generate a required curve, others were point to point, that is to say giving accurate position control as for example to drill holes at specific places on a work piece. These machines were programmed by means of punched tape. It was necessary to develop special ball and roller earings to reduce friction in slideways and rotary movements to give the machine a rapid response time without very powerful actuators. Other machine operations to which numerical control techniques have been applied are punching and riveting, flame-cutting, spark erosion, tube-bending, welding and inspection machines, equipment for the assembly of printed circuit boards and even for cutting out fabric in the clothing industry (A history of numerical control machine tools, M. C. Ferguson, *CME*, September 1978, p. 89). The recent development of hydrostatic slideways has improved accuracy and removed frictional wear on high-pressure tools, although power is required for pumping the pressurised fluid.

(ii) The machining centre. This is a logical development of the old turret head lathe except that here the tool can be changed automatically in the rotating head of a miller so that the machine can be set to carry out several operations on a batch of parts. The British version was the Molins System 24. This consisted of a collection of single purpose numerically controlled machines joined by a computerised conveyor system. The work pieces were palletised by human operators and the pallets fed into the system to pass through each machine in turn. Unfortunately this system was not very flexible since most components could not make simultaneous use of all the NC machines.

(iii) The modern numerical control machine is the CNC or computer numerical control system. Instructions for the processes are fed into a microprocessor with a solid state memory.

(iv) Direct numerical control. This consists of a group of CNC machines linked to a central computer which can feed each machine with the desired program. The computer schedules the work load, provides management information and can determine the optimum sequence for use on the different machine units on a given component. The DNC and CNC systems are being modified for adaptive control in which the sensing of conditions at the tool cutting edge enable optimum feed and speeds to be selected by the computer.

(v) Flexible manufacturing systems. These consist of a set of machines which are fed by means of robots and they can be computer-controlled so that they can be linked in different ways and each one can give a different program. For example a four-station flexible manufacturing system can be run by two or three people where it would require 20 people on the orthodox method. There are perhaps 50 of these in the world and at least half of them are in Japan.

(vi) The unmanned factory.

Many studies have been made of the possibility of having factories in which all the routine manipulation and control are done by a combination of robots, on-line computers, machine tools and automated transfer systems. Fault finding and maintenance would still be done by a small group of maintenance men and steps towards this unmanned factory have been taken, in Japan especially. A diagram of a fully automated factory for flexible production is shown in Fig. 7.15. This is for machining and assembling components and then assembling them into a finished product such as a car or a TV set. By flexible is meant that processes are operated by robots and the product can be carried by either (1) varieties of colour or design stored in the robot memories or (2) human changes of program of the robot memories.

The raw materials enter the factory by lorry or railway truck at A on the left. B represents mobile robots which are instructed by the factory computer O_c to read the labels and take them to the appropriate compartment or bin of the three-dimensional indexed factory stores (C) which informs the factory computer O_c of the quantity in that bin or compartment as it does when materials are taken out so that the O_c has a continuously updated stock record which can also be consulted on a read-out by the human factory supervisor O_H. These mobile robots can move on a network of binary choice paths, for example wires under the floor, white painted stripes on the floor, or overhead or ground rails; O_c knows on which section each one is moving at any given time and ensures that no two ever collide on the same section. The stores also carry spare cutters and other parts for the machine tools, welding rods, gauges, paints and all other needs for operation and maintenance of the subsequent processes.

All subsequent processes are fed with materials and equipment by means of a second set of multiple-choice path, second generation robots (D) moving on the floor, supplemented where there are a number of processes in a fixed sequence by overhead conveyors of conveyor belts. Materials and components then pass through or by-pass various stages, grouped in Fig. 7.15 as hot processor and pressing (e.g. car body sections) E, the machining centres F, finishing and component assembly G. Each of these processes is served by immobile second generation robots with simple sensing of misplacement, tool wear or certain common errors of incoming materials.

At no stage is it necessary to locate and orientate parts from a random pile in a bin or randomly fallen onto a belt because at every stage the component is correctly palletised or placed the right way in a slot feeder as it comes from the previous process, e.g. from die casting, or an automatic lathe. Component assembly would be done by a combination of reprogrammable second generation robots, resettable limited sequence devices and devices such as bowl feeders that require new components for each different shape of piece. These would be used for example if the same screws were used on the assembly of different objects.

From finishing and component assembly the components pass by an automatic conveyor line to the component quality inspection (H) which is carried out by second generation robots and then to final assembly to the complete product (I) and overall quality inspection (J). When the quality inspection robots (H and J) detect faults they reject the product or component and automatically inform both the human factory supervisor O_H with a visual screen and printout and the factory central computer superviser O_c. The human factory supervisor takes two kinds of action (1) he informs the human maintenance group N and tells them to locate the machine or robot causing the error and alter it or replace it (2) he may decide that the error is due to a faulty design of the component or of a tool in the factory and in this case he notifies the human designers P_H to make suitable modifications to the design. This they do in full conjunction with the design computer P_c which has computer-aided design (CAD) and actual computer design (CD) as well as containing in store all the design data of all products and of all machine tools and factory layout.

Products which pass the overall quality inspection are packed or crated by second generation immobile robots J and go to finished stores K. They are taken by mobile second generation robots to where they go by human driven lorries or trains to the wholesaler or retailer and thence to the customer R.

From the customer or potential customers come orders for existing products, complaints and occasionally suggestions for new products. In an organisation large enough to operate a fully robotic factory there would be a Head Office Q_H which would receive this customer feedback. This would be assisted by a Head Office computer Q_c which would store all data on products sold, complaints, proposals and provide statistical read-outs for the human directorate Q_H. The human directorate Q_H would issue policy decisions to the human designer P_H and the human factory supervisor O_H. The computers O_c and P_c would feed back to the head office computer Q_c and call on it for administrative data needed for their work. The Head Office computer would order the raw materials required for each product decided upon.

As long as orders can be obtained for large numbers of complex products such as cars or TV such a factory can be economically justified if the teething troubles and maintenance frequency can be reduced sufficiently. It can reduce ground space, labour costs and eventually customer complaints, although the capital cost will probably be higher. However, as discussed in Chapter 9, a long-term view of the future when, for example, cars are built to last a lifetime to save manufacturing energy consumption does not lead in this direction, nor does the humane economic future proposed by Schumacher in which small firms are more beautiful than large ones.

There is another type of factory where the process can be likened much more to a steadily flowing stream of more or less identical material whose specification changes only occasionally, and where the problems of automation and robot operation are rather different.

Examples of such processes are steel, paper, soap, oil refining, cement and various heavy chemicals. The whole works is divided into separate groups interconnected by the flow of liquids, solids and electricity, steam and heat and each group of processes has its local controllers (level 1 control) which are normally direct digital control and sequential control which takes a batch through a sequence of processes. These stages are already fully automated in many industries and the flow of materials from one group of processes to another is also automated. The level-2 supervisory control and the level-3 management control could be at least partially done by computers. Microprocessors are already used at level-1. These industries do not have much use for robots.

7.8 GENERAL PRINCIPLES OF INDUSTRIAL ROBOT DESIGN AND USE

7.8.1 Accident prevention
Asimov's 'Three Laws of Robotics' were formulated as follows:

(1) A robot must not harm a human being, nor through inaction allow one to come to harm.
(2) A robot must always obey human beings, unless that is in conflict with the first law.
(3) A robot must protect itself from harm, unless that is in conflict with the first or second laws.

These are not laws in the scientific sense, but principles which robot designers should follow, based on the clearly correct ethical principle that human beings are of an infinitely higher quality or value than artifacts made by humans, however sophisticated. Living animals should also be regarded as of higher value than artifacts even if we regard them as of considerably lower value than humans. The belief in the existence of such a scale of values of beings from a stone up to a superman, saint, and even beyond, is indeed one of man's most distinctive qualities.

In fact, however, these principles would only apply to robots which had the possibility of foreseeing the results of various courses of action which they could take and this level of sophistication would correspond to a generation-4 robot. It follows from the previous arguments that it is only necessary to concern ourselves with the design principles for generation-2 robots, although there is no doubt that if the designers of weapons could see a use for robots in warfare they would spend enough skill and money to develop practicable and reliable generation-3 robots. Applications in space studies which have a very close relation to war purposes (surveillance and warhead transport) have already led to some of the most advanced robot studies; but in these cases research funds and capital cost of the equipment are almost unlimited although reliability must be very high and weight very low.

For generation-1 and -2 robots these design principles can be formulated *'the robot designer must try to foresee all the ways in which a robot could harm*

*a human being or damage property and take all possible precautions to avoid
these kinds of accidents'.*

An immobile robot can swing its arm round in a dangerous sweep when it
completes a long row and starts the next, e.g. charging hot glass bottles from a
blowing machine into an annealing lehr. It could be programmed to start the
next row from the end where it finished the previous one but this of course
requires a program twice as long. I have seen a car in very hot weather suddenly
switch on its starter and, because it had been left in gear, drive forward into a
plate-glass window. Similar accidents could happen to a mobile robot because
the power system is controlled by an electronic circuit which can misbehave.
The precautions include such possibilities as

(1) Putting a system of immobile robots into an enclosure which can only be
    entered when the power to the robots is switched off or at the least the
    enclosure is surrounded by a barrier with 'keep out' notices.
(2) Having accessible red buttons which can be touched by a human to immobilise
    the robots.
(3) With mobile robots either keeping humans out of the working area altogether
    or providing them with alternative paths which cannot be traversed by robots.
(4) Surrounding a mobile robot with a light contact pressure switch which
    applies brakes instantaneously or even causes a limited recoil like the rubber
    buffers on a lift door which cause it to open again if it shuts on a person
    or object.

7.8.2 The balance of sophistication
So far most robot applications have been made to replace human operators with
existing factories and designs of objects being made, or to increase the flexibility
of fully automated systems. The aim therefore has been to develop the software
of generation-2 robots to a sufficient level of sophistication and to program
these robots to achieve the desired aim.

Heginbotham has developed a method for designing the layout of a system
using a robot to transfer objects in an industrial situation such as from a moving
belt to a particular machine by means of computer graphic simulation (see
Fig. 7.16) (*3rd CISM-IFToMM Symposium,* p. 434, 'Analysis of industrial robot
behaviour', W. B. Heginbotham, M. Dooner, D. N. Kennedy). The computer is
programmed with a three-dimensional picture of the robot in a discrete number
of positions and with the joint velocity and acceleration between these positions.
It is then fed with the requirements as far as the tasks are concerned and it can
be seen carrying out the task on the graphics terminal of the computer. In this
way the real time required for various possible arrangements can be compared
and the optimum one ascertained.

A similar computer graphics planning for industrial robots has been given
by Warnecke *et al.* (*3rd CISM-IFToMM Symposium,* p. 521, H. J. Warnecke,
R. D. Schraft, U. S. Streier). They conclude that material handling in 35% of all

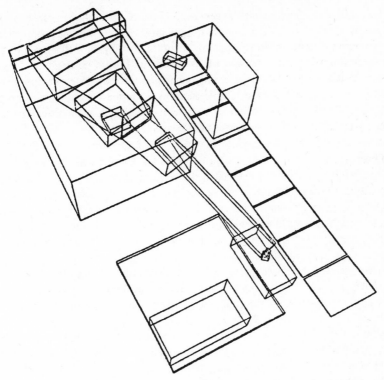

Fig. 7.16 — Simulated Unimate 2000 industrial robot carrying out a palletising sequence.

work places in small and medium batch production could be done by an industrial robot if the chucking, orientating and controlling can also be automated. They are mainly concerned with the two-dimensional planning of the floor layout and they reckon that the best layouts are radial on one of two concentric circles, parallel in line or parallel in two lines.

The possibility of transferring an object from its manufacturing process directly into a pallet or slot feeder with the correct orientation by using a first generation robot, or pick-and-place, limited sequence device, instead of requiring a generation-2 or even -3 robot to locate and orientate randomly arranged objects in bins, has already been mentioned in connection with the robot factory. This is clearly a line that must be developed if robots are to replace humans reliably in many more applications, since the highly sophisticated task of picking out a component from a bin or off a belt will always be relatively prone to breakdown and error.

The other line that must be developed if humans are to be extensively replaced by robots is to redesign the components for the purpose of robot assembly and tool use. Lohr (*Electronics and Power,* June 1979, p. 401) suggests that, in order to assemble small electrical products automatically, with the minimum of technical effort, it should be possible to assemble a product in layers with simple vertical or horizontal movements and that the components should be designed for easy manipulation with the relatively clumsy hands of a robot. One example is that threading a screw into a tapped hole requires a rounded shoulder or a conical thread end. Every engineer knows the difficulty of getting a fine thread to start, especially in a place where one cannot see it.

Screws should be designed so that they can be handled automatically and far fewer varieties used: the choice of fastening should wherever possible be chosen for use by a robot rather than a skilled human. Similarly a standardization of part dimensions to discrete sizes, say, in $\frac{1}{2}$-mm steps can help.

At present it is usual to provide the robot with a range of hands and tool sockets and use a different one for each task. The possibility of increasing the sophistication of the robot hand must be balanced against changing the design so that fewer hands suffice.

There is also the choice between a more versatile robot and a cheaper one with fewer possibilities. At the present state of development it is necessary for the intending user to make the following decisions before ordering.

(1) Limited-sequence device or generation-1 robot (senseless but programmed), or generation-2 robot (with built-in sensory response to touch or force)?
(2) Mobile or immobile robot (only generation-2).
(3) Working space volume and maximum reach.
(4) PPP or RRR for hand placement.
(5) How many degrees of freedom for hand rotation (1, 2 or 3) and whether more than three, i.e. redundant hand placement.
(6) Point-to-point, continuous path or controlled path hand movement. Accuracy and repeatability of positioning required (usually about 0.3 mm is available).

7.8.3 Reliability

A basic law of engineering is: 'the more complex, sophisticated or fast-moving an engineering hardware device is, and the more components and movements it has, the more likely it is to go wrong and either cease to function or function quite wrongly'. Corollaries of this law are the many forms of Sod's Law such as: 'it will fail always when in front of an audience or the Managing Director', or 'it will always find a new and unexpected way of breaking down'. The designer must therefore always try to design with as simple a hardware system as possible which will just do the specified tasks — for example, if all the tasks require only a hand which has a vertical axis, then a PPP movement is simpler than an RRR with chains and sprockets or parallelogram linkages. On the other hand P movements are more difficult to lubricate and keep clean than R movements and give a smaller ratio of maximum length to minimum when operated by direct hydraulic of pneumatic cylinders. The first commercial robot (the Unimate) used a mixture of P and R for this reason.

When he has designed a robot the designer must next try to envisage all the ways it can break down or go wrong (starting of course with the dangers to humans discussed in 7.8.1) and then try to improve his design to reduce the likelihood of each of these.

However, every experienced designer knows that he will find in practice consequences of Sod's Law, namely that it will break down in ways that never occurred to him. If he has had to get his design past a hostile committee they will have thought of lots of other defects in the design but these will also rarely coincide with the ones revealed by empirical trials! Even if the design is taken through the single prototype and multiple prototype stages (described in Chapter 9 of *How to Invent* by M. W. Thring and E. Laithwaite (Macmillan, 1977)), it will find new causes of failure in extended use in the tough conditions of the factory floor.

Engelberger ('Designing robots for industrial environments, *Mechanism and Machine Theory,* **12**, p. 403; *Robotics in Practice,* p. 82) who has had more than 20 years and more than 4 million hours of industrial robot experience has been able to design out so many of the causes of downtime that he has produced a 98% field reliability under working conditions by a mean time between failures (MTBF) of 400 hours and an average downtime for repair per incident of 8 hours. When a higher reliability is needed it is necessary to have a back-up of a relief human crew or spare robots. He describes a reliability assessment in which the individual part failure rates (in number of failures in 10^6 hours) are added together with the subsystem (non-part) failure rates. 10^6 hours divided by the total failure rate gives the estimated MTBF. It is interesting that both in terms of part failure rate and sub-systems the electrical/electronic components are more reliable (1.2 and 1.8 times respectively) than the mechanical, hydraulic ones. The total part failure rate is 1.7 times greater than the sub-system failure rate caused by tolerance build-up, critical interface tolerances, customer abuse, unforeseen environmental problems.

As the result of his experience Engelberger recommends that there should be no electrics or servos on the robot's hand because it has to enter hot hostile places, in other words it should be operated by a mechanical linkage such as a cable, sliding rod or rotating shaft from further down the arm. For working in extreme shock conditions and in corrosive or explosive atmospheres the electronics should be mounted separately, while in potentially explosive atmospheres, such as in paint rooms, the power supply should also be separated from the arm. The robot's skin should be non-flammable, and rotating or sliding joints should be booted with flexible covers to protect from abrasive dust: in principle a rotating joint is easier to protect than a sliding joint which requires a concertina. He recommends that the drive trains should use hardened gears and be pressurised to reduce infiltration of dust and dirt. The robot logic design should be heavily protected from fluctuations in electric power input and from noise pick-up from surrounding wires to the robot's communication links.

7.9 OTHER POTENTIAL ROBOT FUNCTIONS

As well as in the factory, robots could be (and have been in some cases) used in:

(1) Offices.
(2) Hospitals.
(3) Warehouses and central stores.
(4) Homes: for routine domestic tasks.
(5) Dust carts; garbage collection. Concourse floor sweeping.
(6) Farms.
(7) Fires in warehouses, factories, hotels, houses.

7.9.1 Offices, hospitals and warehouses

Both in offices and hospitals prototype installations have been made of mobile distribution robots that can be instructed to take goods or packages from point A on one floor of a large building to point B on another floor. They can operate the lift and enter rooms following a multiple-choice path system. The distribution of hot food in a hospital and tea cups and files in an office have been considered. In the latter case electronic reproduction on a TV screen of material stored on the central computer data bank will probably replace much of the routine file distribution although the conveience of rifling through an accurately kept file, dealing with a subject rarely used for several years, is difficult to reproduce electronically, just as finding information in a printed encyclopaedia is astonishingly effective after one has learnt to use it, because the human mind can readily think of several different headings under which to look. Warehouses and central stores can also use a similar mobile robot.

All these types of mobile robot can be programmed to follow wires under the floor, or lines painted on the floor, or to move along at a constant distance from a wall using sonar or another proximity sensor. When they come to a branching decision point their microprocessor takes the decision according to the instructions given to it before it sets out on its journey. The decision point will supply a coded signal like a lighthouse so that the robot knows it has reached the numbered decision point (as was done mechanically 20 years ago in the 'rat in a maze' of Fig. 7.3). Dead reckoning (that is calculation of the distance moved in X and Y coordinates by counting the wheel rotations between each change of oreientation — the latter being known by a gyro compass or by counting differential rotation of a pair of wheels on an axis perpendicular to the direction of movement) is always inaccurate and the inaccuracy can become very considerable as the distance travelled increases, because of wheel slip or changes in wheel radius due to changes in tyre pressure or tyre wear. Hence these mobile robots must always be fitted with position-sensing devices accurate to a few millimetres for final positioning which can correct the position both during the journey and especially at the end. This may be done by the underground wire or painted line detected by the robot's sensors but it will also need a position-sensing device, tactile, visual, sonar or radar to ensure that the final position is correct in relation to the pick up or discharge point.

If the transport robot is to pick up the object, whether file, package, or small or large component from a bin, shelf or delivering lorry, and is to unload it to a shelf in a 3-dimensional store or to an assembly belt, machine tool or immobile robot then it will require a generation-2 or even generation-3 robot arm and hand with sensors to locate the object and read its label or the bin label (which can printed in any suitable robot language).

7.9.2 The domestic robot

Fifteen years ago I was working seriously on the idea of a domestic robot to do all the routine housework ('Towards the domestic robot', M. W. Thring, *Discovery,* March 1964, p. 26; 'Domestic automation', *Trans. Soc. Instrument Technology,* June 1964, p. 47; M. W. Thring and D. F. Nettell, 'Automation in the Home', *Electronics and Power,* Nov. 1968, p. 440; I robot per la casa e l'industria, *Mondadori Enciclopedia della Scienza e della Technica,* 1968, p. 238; 'The domestic revolution', *Jl. Royal Society of Arts,* 1963, p. 556; 'A robot about the house', *New Scientist,* 2 April, 1964, p. 19). The idea was that a robot could be developed which could do such jobs as vacuum-cleaning the floors, or even scrubbing the kitchen floor, dusting, laying and clearing the table, loading the dishwasher, going upstairs and making the beds, washing and ironing and all the other tasks that are done regularly every day or every week in the house. The wooden model of such a conceptual robot, shown in Fig. 7.17, has

Fig. 7.17 – Domestic robot model.

one long arm with a large parallel jaw gripper at the end which can reach up to the ceiling or down to the floor. It also has a generation-3 visual system at the top of the pillar that carries the arm. The base is a tray for carrying tools and trays, under which is the body with rechargeable batteries, drive motors and microprocessor with a range of stored pre-instructed programs on discs or tapes. The body is carried on four of the rimless stairclimbing wheels described in Chapter 6. A second hand with three fingers for gripping round objects sticks up from the body so that an object can be placed in it such as a saucepan by the other arm and rotated slowly or rapidly about its axis while the other one polishes or scours it.

At that time I was concerned with the problem of the highly educated wife who spends so much time on the household chores. Since then, however, (*New Scientist*, 19 Nov. 1981, **92**, p. 500) as described in Chapter 9, I have gradually come to realise that this problem is not worth the engineering effort needed to solve it, even if governments or industry were prepared to divert the necessarily large effort to it from weaponry, the space race or nuclear power, for the following reasons.

(1) I have found by personal experience that doing such tasks as well as one can is a therapeutic change to balance one's over-intellectual activities.
(2) If mankind continues to spend such a large part of our engineering effort on weaponry then we shall have World War III with total destruction of civilis-ation long before the domestic robot is available.
(3) The problems of desert growth, poverty, starvation and disease of the larger part of the world's population are so much more urgent than this one.
(4) Rising unemployment in the developed countries must inevitably cause a total rethink in our attitude to robots.

7.9.3 Cleaning and sweeping
A few years ago I was told of a problem in an oil-rich Arabian country where the airport concourse required regular sweeping, particularly for cigarette ends and there was a shortage of labour to do this. This could certainly be done by a mobile generation-3 robot with suitable dustpan and brush or vacuum sweeper which could cover the floor with a fixed program to avoid pillars and seats but also a sensitive tactile or sonar device to enable it to go round people and luggage. A similar robot could sweep a factory floor at night.

The future of garbage collection or dustbin emptying will also have to be very different from the present when we move from the perpetual growth economy to the equilibrium economy. The need to recycle everything possible will require the householder to sort domestic refuse into various separate bins such as compostable organic refuse, broken glass, metals (e.g. cans, batteries), hardcore, combustible material, paper for recycling and the dustcart will have to keep these separate. The use of a single bin with many compartments which can be automatically inverted over the correct chutes on the dustcart could well be carried out by a robot of strength and arm greater than that of a man, but more likely this would be a telechir and the operator would walk beside it and control its path and handling operations.

7.9.4 Farms

The use of robots for farm operations such as covering a whole field with a plough, sprayer or combine harvester or stacking large bales of hay or straw, has been considered by various research organisations. Tractor driving for these field operations is a very skilled task for a human to cover the ground completely with minimum overlap. It is theoretically possible to do it with a robot driver by a combination of rough guidance from radar points around the perimeter of the field and fine guidance by a visual or other sensor close to the ground in front which detects the edge of the previous row.

While it is true that modern ploughing is somewhat less interesting and skilful than the old horse ploughing it is perhaps 20 times faster and uses up to 50 times as much power; thus the cost of manpower is no longer the major cost and the developments of robots for this purpose is undoubtedly beyond the optimum point of sophistication. However, fairly complicated automation systems are already being introduced in modern farming, for example a laser beam is set up a few metres above the ground at the required angle and slope so that the tractor inserting a land drain can automatically insert the drain at this slope by keeping the drain the same total distance below the beam using a beam sensor at the top of an arm. Another application is the microprocessor which keeps the rate of supply of liquid fertiliser per unit distance constant while the tractor towing the sprayer runs at variable speed at the wishes of the operator. This machine can even vary the amount of feed to the two sides of the sprayer as it goes round a corner to keep the coverage per unit area constant.

7.9.5 Fire detection and extinction

One area where robots (and telechirs) can clearly be of immense value is the detection and extinction of accidental fires. Rescuing people from a burning building will probably require a water-cooled telechir with a human fireman outside the building using it to search and carry a cooled rescue chamber with oxygen mask into which a human can climb, or be lifted if unconscious. However, designs have already been made for a robot nightwatchman which would follow a closed path round a deserted factory, office or warehouse and detect fires. In one design the robot travelled on an overhead rail through small holes with spring closed doors high in the walls between rooms. It carried an infra-red scanner which rotated slowly about a horizontal axis coinciding with the direction of movement so that the scanner covered all the surfaces of the room including the ceiling which were visible from its path, in a helical scan with overlapping edges. If the sensitive scanner detected a source of heat (all intentional sources would be screened from it) it would stop the robot movement and its rotation and stay pointing at the heat source. If a second fire detector, such as an ion detector, confirmed that there was a fire, then a fire extinguisher jet would be directed along the scanner line and an alarm would sound in the human night-watchman's office. The infra-red scanner can also be used to detect the warmth of the body of a human intruder.

Figure 7.18 shows a demonstration robot firefighter built in 1962. It followed a track plotted on the table, detecting the track by photocells, and determining its direction by the gyro compass at the back. The distance travelled was determined by counting the revolutions of the back wheels and the drive and steering were by the front wheel which could be rotated by ± 90° from straight. An arm sticking out in front carried a bimetallic switch and there was a photocell detector on a headlight fixed to the front wheel carriage. This carriage oscillated during the steering process and if the photocell caught sight of a flame the robot left its track, homed on the flame, and when the bimetallic switch detected the heat of the flame the robot stopped and brought the fire extinguisher nozzles onto the flame. This first prototype was liable to chase the sun!

Some consideration has been given to robots for underwater work but in general it is better to combine the local artificial intelligence of the robot with supervision by humans on the surface of the water in a telechiric system of the kind discussed in the next chapter. In the case of work on a remote planet or in space, however, where the time for sending of information from the space device to the human in the control tower on arrival and for returning his adaptation to the signal may be many seconds or even minutes, it is essential to have a robot on the spot for dealing with certain emergencies such as sensing a crevasse in front.

7.10 ARTIFICIAL INTELLIGENCE FOR ROBOTS

Work on artificial intelligence has been carried out for a number of different reasons.

(1) To increase our understanding of the working of animal and human brains by producing artifacts which could produce the same response to external circumstances as those of the biological system.
(2) To deal with strategic and planning situations which require the interaction of too many factors to be comprehensively handled by a human planner; much of the work on games-playing computers has been with this aim.
(3) Only the third aim is relevant to the practical study of robots and that is to enable a robot to do some of the jobs that a human worker can do with his trained brain, hands and eyes.

In order to discuss artificial intelligence for robots it is necessary to try to understand a little about human intelligence. In Chapter 1 I have already mentioned the Eastern idea, developed and applied in *All and Everything* by G. I. Gurdjieff, that man has three brains, while in Chapter 2 I have tried to summarise

Fig. 7.18 – Robot firefighter.

some part of modern knowledge of man's neural system. The concept of man's three brains can be verified by direct self-observation. They are:

(1) Intellectual: dealing with ideas, concepts, abstractions, theories and models of reality. The intellect works by a slow sensation process.

(2) Emotional: dealing with emotions or feelings including both the negative emotions such as fear, worry, depression and anger, and the more positive ones such as religious aspirations, long-term motivation for life, love and caring for others, a sense of values and scale of values for the whole universe and for different levels of being at different times in oneself and between individuals who have developed themselves to different degrees, happiness and enjoyment of life. The emotional brain can work with incredible speed as when one sees someone for the first time.

(3) Physical: the body brain. This, like the emotional brain, works very much faster than the intellectual brain because it can coordinate simultaneously dozens of muscular movements into a single overall action which is continuously adapted to external sensations such as the sight of the ground in front of one, the estimation of weight of the ball thrown for me to catch, or the slipperiness and rigidity of the ground as one puts one's foot down. Even more marvellous is the way the body brain can coordinate millions of optic nerve signals into a single picture in a flash. Sometimes one's eye catches a glimpse and produces a wrong image from one's vast range of stored memories typified by Lewis Carroll's

> I thought I saw a hippopotamus
> I looked again and saw it was a
> lady getting off a bus.

However, the car driver who catches sight with the corner of his eye of a child running into the road and immediately stamps on the brake is an example of the rapidity of the simultaneous working of the emotional brain which produces a strong feeling that one *must not* hurt the child and the physical brain which produces the vision of the danger and works the necessary muscles to avoid it.

Equally wonderful is the ability of the body brain to store whole languages, not only the vocabulary but also the syntax and the change of meaning by changes of intonation of the speaker as when the same words can be a statement or a question. When you are learning a new language, or when learning to drive a car, it works through the slow processes of the intellect translating word by word but sufficient practice can bring it into the speedy body brain and then 'you think in that language'. Similarly when learning to play a musical instrument, use a new tool or even a breakfall in Judo, one starts with the intellectual brain which is much too slow and only after hours of practice does one get it into the rapidity of the rapid, co-ordinated, simultaneous, muscular movements of the body brain.

The natural sense of rhythm also belongs to the body brain and the ability to recognise accurately musical note frequencies and dozens of different shades of colour or tastes are all education of the body brain.

Thus we see that each of the three brains in man has its own method and speed of operation entirely different from the others and its own memory storage system which has to be educated by a different kind of effort.

I have postulated in Chapter 1 that a human artefact can never contain an emotional brain so that a robot can never experience human feelings positive or negative. The only way it can deviate from its instructions will be accidental errors, although it can of course be instructed to carry out random activities. I will therefore define artificial intelligence for a robot as *the extent to which it can reproduce the fully trained activities of the human intellectual and physical brains.*

The computer can move towards the intellectual activities of a man, together with the body brain activity of language, while the robot can aim to add the body brain activities; although clearly language nuances, visual pattern recognition to the range of a human's, and coordination of bodily movements as in running are at present far beyond the capacity of the most advanced computer-brained robots. Indeed I believe   always be outside the range of economic multiple production   will they require a neural system as complex as ours (w   each connected by a trainable connections to $10^2 - 10^3$ others   have to be trained for as many hours as we are trained for such activities. However, sheer complexity and man-hours of training time do not constitute an insuperable obstacle for laboratory researchers who have worked on similar training problems for monkeys, so we may see a few such robots operating in laboratories if our civilisation avoids World War III for long enough.

There have been many other definitions of artificial intelligence. Russian workers have defined it (A. N. Radchenko and E. I. Yurevich, 'On definition of artificial intelligence', *1st CISM-IFToMM Symposium,* vol. 2, p. 91) as the ability to have adequate behaviour in a changing environment, and they have suggested that the level of the intelligence may be measured by the exactness and brevity of the translation between the afferent text (that is the sensations coming into the nervous system from the outside world) and the efferent text (that is the control of the movement of the system). In the human system the afferent text is brought in by four million optic and acoustic nerve fibres while the muscular control is conveyed by 3.4×10^5 fibres. Other definitions of artificial intelligence in a robot are that humans interpret its behaviour as trying to achieve goals which satisfy drives analogous to human hunger and thirst (*The Robots are Coming,* (ed. F. H. George and J. D. Humphries), NCC Publications, p. 72) and the ability of the intelligent artefact itself to synthesis and/or use and revise some kind of internal model that enables a desired type of action and inter-action with the environment (*The Robots are Coming,* p. 135) to be achieved.

Raphael (*The Thinking Computer,* W. H. Freeman, 1976) describes a computer as a 'big, fast, general purpose symbol manipulating machine' which can be developed to be 'flexible, decision-making, problem solving, perceiving' (p. 5). Lady Lovelace wrote in 1842 'the Analytical Engine has no pretensions to originate anything. It can do whatever we know how to order it to perform', and this is still ture of the most effective modern computers and hence also of robots with computers for instructing their movements.

So far improvements to computers up to the fourth generation computer ('The race for the thinking machine', P. Marsh, *New Scientist,* 8 July 1982, p. 85) have been to increase the speed, fast and slow storage capacity and number of binary switches by many orders of 10 and improve the reliability and reduce the size enormously; but the machine has remained a system for handling binary numbers by a single path at a time with many programmable points of choice between alternatives. They can handle 10^7 units of data per second but one at a time. The fifth generation being studied in several countries will have a number of parallel mechanisms for sending instructions from the memory and processing them in parallel paths. This will still be far short of the human ability to combine a two-dimensional optical image with estimates of range and orientation and the memory store of what things look like, or the human language understanding. The fifth generation computer will operate on languages that are easier for the human operator to learn and it is even hoped to be able to feed into the computer's memory some of the things we have learnt about reality from our senses as children or teach it to find them out by its own observations.

Levels of intelligence
One of the difficulties of defining artificial intelligence is that there is a whole spectrum of different levels of artificial intelligence just as there is a whole spectrum of human intelligence, ranging from the moron to Einstein. I shall try to make this more precise by defining four levels of artificial intelligence for robots.

(1) The purely mechanical system which does exactly what it has been programmed to do with only minor alterations in its standard behaviour corresponding to simple external changes which, for example, operate a sensory switch. This corresponds to Generation-2 robots. Examples of this lowest level of artificial intelligence are the various devices I constructed many years ago using a Post Office uni-selector switch as the brain. These include the table-clearing robot and the mechanical rat in a maze which explored eight different paths in order until it came into the one with a piece of 'cheese' in it and continued to follow this particular track until the 'cheese' was removed. The automatic telephone based on uni-selector switches, automatic assembly robots fed by bowl feeders with jigs which have to be changed for each new job, the original cam-operated lathe and numerically controlled machine tools, limited sequence devices and Generation-1 robots

can none of them be said to have any artificial intelligence since they have no sensory adaptability at all.

(2) *Algorithmic intelligence.* Robots of this type can be given complex programs which at every stage can be told 'if the result of the previous stages or external circumstances are observed to give result "A" then proceed to action one. If result "A" is not obtained then proceed to action two'. In the case of computers this frequently takes the form 'if a certain numerical result turns out to be less than a present figure then you proceed to the next step. If it is greater than this figure then you repeat the previous operation or proceed to a different step'. In the case of a robot it might take the form 'if you run up against an obstacle then go round it and return to the previous path'. A more elaborate example is opening a combination lock by trying all the 10,000 possibilities until the right one is found. Robots with algorithmic intelligence are the third generation robots which are now being built and tested. They can be given senses and can do sequential pattern recognition because digital computers, although extremely fast, carry out a single sequential series of operations; the TV camera for example is a sequential pattern monitor while algorithmic computers can learn to play games in which there are sufficiently few moves that all the possible consequences of a given move can be worked out or in which they can be given a numerical evaluation function which is so perfect that the improvement of the situation at the end of the number of moves that can be calculated ahead resulting from every possible move can be calculated so that the best move can be chosen.

The most famous example of an algorithmic robot is the SRI 'Shakey' (*The Thinking Computer,* B. Raphael, p. 275). This robot had a TV camera and touch and distance sensors and could wheel itself about in a laboratory environment of interconnected rooms with large wooden blocks in it. The TV picture from the robot could be processed in a PDP 10 computer with 200,000 36-bit words of fast core memory. The program was on four levels: low level action for movements; intermediate level actions for groups of movements, such as pushing or 'go to', and for processing picture data of the simple blocks; the third level was a planning mechanism to construct sequences of movements groups; and the fourth level was the executive which invoked and monitored the whole task. As it had no arms or hands its actions were limited to pushing the boxes around or selecting a particular one and putting it against another one.

Algorithmic intelligence is limited to carrying out tasks with procedures which always work. Real situations often contain unexpected obstacles and procedures do not always work in them.

(3) *Heuristic intelligence.* All situations to do with the future, all situations which may contain unexpected intrusions from outside and all problems which contain too many possibilities to be worked out in detail to the

conclusion have to be dealt with at the heuristic level. We humans are constantly making plans to deal with incomplete intelligence. As a simple example we estimate how much time to allow for fluctuations in inter-mediate transport time to catch a train at a main line station so that the chance of missing it is small but we do not waste too much time by coming too early. We make a strategy to seek our goal and this strategy may have a hierarchy of sub-goals. Ideally what is required is called 'fuzzy logic' of which logic is the special case when the fuzziness is reduced to zero. However, digital computers cannot really cope with fuzzy logic and the way the chess-playing problem has been solved so far is by giving the computer an evaluation function which assesses the value of different pieces and of different positions so that all the possibilities after the three or four moves which can be calculated ahead can be evaluated. In the more complex strategies dealing with social or warlike problems all that the artificial intelligence can do is to estimate probabilities of various results. Heuristic pattern recognition implies recognising the pattern of a three-dimensional object at all possible projection angles and distances and implies a guess at recognition to enable it to put the pattern into a certain category or not. Parallel pattern recognition which takes simultaneous account of a group of a large number of points can be used here but artifacts are still far from the comparative success with which a human being can look at the face or photograph of another human being and say whether it is a man or a woman. Even we are sometimes wrong!

Heuristic intelligence involves true learning or concept learning in which a new knowledge structure can be built out of previous knowledge and sense perceptions of more elementary concepts. It can form a picture or a model or a symbolic network representing the meaning of words.

While the development of heuristic intelligence in computers is of very great interest and of great value in using computers for sociological decisions it tends to be much slower than the intuitive decisions of a human being and it requires a very great level of computer power. Thus in almost all situations where complex physical manipulation is required on this level of intelligence it will be better to use the intelligence of a trained human by means of the telechiric systems to be discussed in the next chapter. However, there is one exception and this is in space applications where the distance is so great that the round trip of information back and instructions forward for movement would be too slow to deal with emergencies. In this case, therefore, it is desirable to install heuristic intelligence in an artifact, but it must be fast.

(4) *Creative intelligence.* When human beings have to do a certain task many times over which requires the use of tools and skill they can achieve a greatly improved level of performance by using their heads to work out better ways of doing it and find or develop better tools. An elementary example is planting dozens of daffodil bulbs and finding that the use of a

spade to take out a square turf and plant four or five bulbs at a time is quicker and better than using a trowel and planting them one at a time. This steady working out of an improved technique, which requires a knowledge of the different tools available and the experience of how to use them, is quite a creative activity which would not be worth programming into a robot because this kind of job can be much more satisfactory for a human to do.

When we come to real originality, however, not only is the human intellect involved as well as the craft skills but also it involves the emotional brain of a human being which I believe can never be put into an artifact. Inventiveness has been described as lateral thinking by de Bono and as bisociation by Koestler. I have described it as leaping over the fence into a new field in the book *How to Invent*[†]. It is the truly original creativity that can never be programmed because it brings in a solution which is just what the situation needed but which has no possible associative or automatic logical connection. This creativity is involved in writing a genuinely original poem, play or novel, in the 'hypothesis-formation' of the scientist which leads to a totally new synthesis of hitherto discordant effects and in the inventive leap of mind of the engineer. A simple example is to say that, however good a game of chess you can make a computer play, it could not wake up one morning bored with chess and say 'I have invented a better game'.

7.11 CRAFTSMANSHIP FOR ROBOTS?

Not nearly so much thought and attention has been paid to the subject of levels of craftsmanship as has been paid to levels of intelligence. However, from the point of view of robots and telechirs it is also important to understand that there is a spectrum of steadily increasing craft skills finishing up with those that could probably never be done by a robot or even a telechir. In the case of the robot the skill is limited both by the limitation of the ability to duplicate the human hand/eye coordination as well as the relative clumsiness of mechanical limbs, whereas in the case of a telechir it is only limited by the clumsiness of mechanical limbs, since the human controller can ultimately exercise his own trained hand/eye coordination. Thus the robot will probably be limited to the first two levels of craftsmanship, whereas the telechir can proceed to the third but never achieve the fourth.

Levels of Craftsmanship

Level 1. These are the tasks which when done by a human require neither mental attention nor highly accurate movements although they do require a certain measure of hand/eye coordination, that is they would be much more

[†]M. W. Thring and E. Laithwaite, Macmillan, 1978.

difficult to do by a blind person. An obvious example is washing-up in a kitchen where one has done it many times before, the mind is not concerned with making decisions since the physical memory knows how to deal with the different types of dirt and where to place the different objects afterwards. Walking along a path one has walked along many times before, is another example where the head can be engaged in an entirely different activity without the physical activity suffering, even if there are rough stones on the path, stiles or stepping stones across a stream.

Level 2.   Involves accurately coordinated movements which require to be learnt but can be learnt in a few minutes or hours. Occasional intellectual decisions are required, such as deciding where to look for a certain object. These tasks would be considerably more difficult for a robot probably because of the greater difficulty of handling the objects, examples are making a bed or folding sheets, packing a suitcase, changing the drill in a chuck using a chuck key, knocking a hole in a wall or driving a car.

Level 3.   This is highly skilled tool handling of the type that takes apprentices many months to learn. The axemanship of a skilled woodman, carpentry with hand tools, wood turning and experienced driving which is aware of the note of the engine, methods of economising fuel or getting through heavy traffic without aggression are examples.

Level 4.   This is the technique of the artist, musician or sculptor which requires not only daily practice for several years but also requires the full involvement of the emotional brain to give to the activity, significant value for other people.

7.12 ROBOT LEARNING

One can similarly distinguish four levels of robot learning.

Level 1: Rote Learning
This implies memorising a table of data which can subsequently be used. Examples of this are the branching tree search method such as can be used by the computer to learn how to play noughts and crosses by storing information of all the possible winning and losing games. A mechanical rat which explores a two-dimensional maze and remembers all the abortive side tracks so that it can avoid them in future is another example. Rote learning can even be applied to the game of draughts by memorising the early positions of the game, together with a static evaluation of this position. This is possible because the number of early positions is very limited for this relatively simple game.

Level 2: Parameter Learning
This implies that the computer gradually works our parameters which describe simple categories of a classification and when they are set up it can then assign

any new samples to the appropriate category. A human may present a number of samples of each category to the computer telling it which category they belong to but the computer determines for itself what are the common features or parameters for each category. Examples are alphabetical letters of different type fonts and electroencephalographs and electrocardiograms. Parameter learning has also been applied for the draught-playing program where the computer gradually improves its evaluation function by giving different weight to different factors until it decides the same moves as an expert or by comparing the position evaluation factor with those obtained by looking a few moves ahead.

Level 3: Method Learning

Here the computer or sensing robot finds the successful solution to a few problems, all examples of the same process, and generalises this solution in the form of an algorithmic or procedural program which can then be applied to a wider range of similar problems. Simple examples are learning the technique for long division or extracting the square root, or the technique for taking things off and putting them on to hooks, which can be applied to very different hooks and very different objects.

Level 4: Concept Learning

This implies building a new knowledge structure out of previously known more elementary concepts. It is probably important if we wish to enable computers to make use of words with an understanding of their meaning. Fairly simple words can be learnt with a set of building blocks by making several arrangements which conform to the concept and several others which do not; but more complex abstract ideas like beauty or goodness are probably prohibited by the lack of emotional brain in a computer. Concept learning is definitely heuristic since it is almost always possible for a clever designer to get round any exact definition by making an object which serves the same function in a new way.

CHAPTER 8

Telechiric machines

8.1 DEFINITIONS AND USES

The word telechirics was first proposed by J. W. Clark of Battelle ('Telechirics for operations in hostile environments', Battelle Report; see also his '*Ocean exploration by machine*', *Sea Frontiers*, **8**, p. 76–83, May 1962). The word telechir means literally 'remote hands'. However, I shall take a more general definition as any system in which a man's brain controls mechanical hands to do complex manipulative tasks by means of signals provided by muscular contractions or nerve signals to muscles of his own body. This broad definition includes systems where a man uses mechanical hands or grippers close to his body to increase his strength, where a person who has lost the use of a hand controls a mechanical hand close to his body by signals such as movements of his eyes, sucking and blowing with his mouth and movements of his shoulder muscles. It also includes all the telechirs which are used remotely to enable a man to work in a hostile inaccessible region and it includes cybernetically operated micro-hands for work too fine for a man's hands. It does not, however, include the remote control of single-purpose machines such as a mole miner which has to be steered through the coal seam, by a human receiving sense impressions (although this device will be considered under the general section on mining in this chapter) nor does it include cranes or forklift trucks since these are purely single-purpose devices without the capability of complex manipulation such as rotation of the object about a horizontal axis.

Figure 8.1 is a diagram of all the feedforward controls and feedback information that are possible in a complete telechir for remote work in an environment requiring mobility. The human operator A sits in a chair at the master station and can use his hands either to move a joystick (J_1) to drive the telechir about in its environment, or other joysticks (J_2) for example to raise or lower the 'head and shoulders' assembly C by a vertical traverse device on the telechir, or he can place them in *control hands* D consisting sometimes of mittens, the fingers going into one slot and the thumb into the other sometimes of a pistol grip with trigger or split cylinder. These *control mittens or control hands* servocontrol the 14 degrees of freedom of the telechir's two hands. Usually this is done by attaching the mittens to *control arms* E which exactly duplicate the slave arms (although not necessarily to the same scale) so that the 14 controls do not have to be unscrambled by computer. These control arms do not have to be duplicates of the human master's arms; for example, they do not have to have the ability to move the elbow round while the hand remains fixed in space

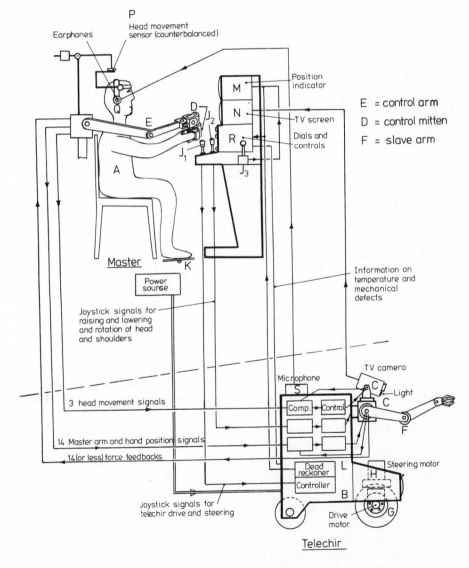

Fig. 8.1 – Diagram of components of a complete, freely mobile, telechiric system.

and orientation unless the telechir needs this freedom for access to different positions. They can thus have any type of movement that is convenient for the telechir in its environment so long as the human operator *can have vision and sensation as if his own hands were doing the job sufficiently so that he can use his trained hand-eye coordinated skill.* Various types of control arms that have been used for telechirs are shown in Fig. 8.2. When his hands are not in the mittens D, the control arms E and the telechirs arms F are in the closed position, unless the operator has locked them holding or carrying some object. He is then free to use his hands to work the joysticks. Movement of the telechir is shown diagrammatically as a system with two free wheels at the back and a wheel at the front driven by motor G which can be rotated by ± 180° from the straight forward position by steering motor H position. This is a drive system with the ability to go rapidly in any direction or spin round the midpoint of the back axle, but if the working environment of the telechir requires climbing over fallen stones or up a flight of stairs then any other walking system as described in Chapter 6 can be used; in space, rockets can be used, and under water, propellers. It is also possible to control the motors H and G by foot pedals K when the operator has his hands in the control mittens.

The dead reckoner L counts the turns of the drive motor G and knows the direction of movement of the body by means of a gyro compass and of the drive wheel by an angle indicator on the steering motor M. This information is used to indicate the position of the telechir on a screen M which carries a map of the working region in which the telechir can move. The operator can make fine corrections to this information by means of Joystick J3 when his vision system (TV cameras C and TV screen N) or his tactile information tells him that a noticeable position error has occurred as a result of cumulative inaccuracies.

The TV cameras C can be a pair to give binocular vision to the eyes of the operator A and a counterbalanced cap or helmet P can be used so that linear movement or rotation of the operator's head can be followed by the cameras C to give him 3-dimensional viewing. Alternatively he can learn to use 2 TV screens, side by side, receiving information from widely spaced cameras on B, and ultimately he will have genuine 3-dimensional vision provided by remote moving laser holography.

The control mechanical arms E moved by the operator's hands not only servocontrol the 14 movements of the telechirs arms and hands F but also receive a limited number of feedback force signals, say, two at the shoulder to enable the operator to sense the weight and inertia of the telechir arm, hand and object held. It can also provide one or more tactile signals to his hands, or these contact signals and others for contact of the telechir body can communicate with dials R or give an audible warning signal to the earphones Q. These earphones can also receive sounds from the microphone S.

The power source is shown as electric, transmitted by cable with the control and feedback wires from the operator's cabin, but provided the communication

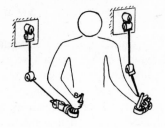

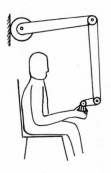

1. Exoskeleton
   (Mosher)

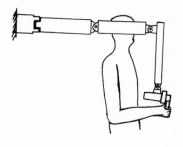

2. Rear Overhead (standing)
   (Wilson)

3. Rear Overhead (seated)
   (Mascot)

4. Front Overhead (seated)
   (Goertz

Point fixed to
control cabin

Fig. 8.2 — Various types of control arms used in telechiric systems.

is available it may be battery-operated or even a local prime mover with its own fuel tank.

The first work on telechiric machines was done in connection with the handling of radioactive materials which are known to have extremely harmful effects on the human body and from which the human body must be shielded, either by a considerable distance or by massive shielding. A great deal of work has also been done on telechirs in space, where it is much cheaper to put a non-air-breathing machine than a human being, although in this application there is the very severe handicap of the long time delay of the two-way radio communication. Active work is now beginning to enable people to do jobs deep in the sea without going down in diving suits or diving bells and work has just been started on the possibility of mining coal and other minerals by means of telechirs. This work will all be described later in this chapter but here we can speculate on how far it will be desirable to develop telechirs in the twenty-first century.

One of the present uses of rather elementary telechirs is bomb disposal where it is obviously far better to put a machine at risk than a human bomb-disposal expert. However, if we have not got rid of the need for the disposal of these devices for random civilan murder by the next century we shall probably have destroyed our whole civilisation in World War III and we need not pursue this line further. It will certainly be worthwhile developing undersea telechirs to the point where it is totally unnecessary for human divers ever to go down and we shall probably be able to operate an oil-drilling rig on the base of the sea with all the necessary work carried out by telechirs (see Fig. 8.3). Similarly there is no question at all that we shall have to develop mining systems whereby any mineral, but especially coal, can be won without men ever going down the pit.

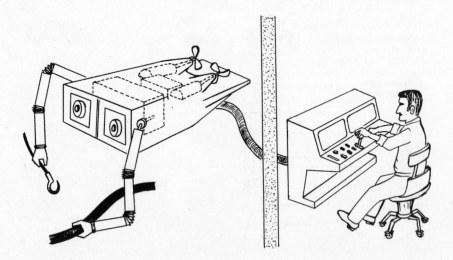

Fig. 8.3 – Diagram of underwater telechir.

This will give us the possibility of mining very thin seams without having to make roadways sufficient to carry men and without ventilating the mine. It also means that we can win coal far under the sea and at depths which are intolerable for men.

We shall still have certain tasks in factories which involve the risk of explosion or the escape of poisonous gases and there is no doubt that, once we have solved the problem of a relatively cheap telechir, all this kind of work can be done remotely from a comfortable and safe control cabin. Other ways in which the development of telechirs are likely to cause considerable improvements in human life are: (1) the use of a telechir for going into a burning building and rescuing people or extinguishing the fire chemically and safely from inside instead of playing hoses on it from a considerable distance; similarly (2) an empty building can be controlled at night for fire or other hazards with a single nightwatchman watching a television screen for the telechirs.

However, there will always be some jobs where the skill of a telechir cannot match the skill of the craftsman's hands handling the tools directly, quite apart from the fact that a telechir will always be slower than a human working directly with his hands. Examples are the craft skills of wood-carving or painting a picture, jobs requiring very fine sensory feel combined with close vision, such as fitting a small bolt into a nut or avoiding cross-threading a very fine thread, and jobs where it is necessary to handle materials with complex dynamic properties, such as folding a sheet or tying a shoelace.

Apart from remoteness there are other advantages in telechirs compared with the direct use of the human's arms and hands:

(1) Mechanical arms and hands can be scaled up by a large factor, both in length where arms up to 15 m long have been used and in strength where arms have been developed for lifting half a ton. The force feedback is correspondingly scaled down so that the operator feels as if he were handling a light object suitable for his strength; equally mechanical hands can be scaled down to perhaps one-tenth of the size of the human's hands associated with 10-fold visual magnification and an amplified force feedback[†]. It is probably not possible to go much further than one-tenth because the increased magnification gives a very short depth of focus, and depth of focus is necessary for three-dimensional tasks.

(2) It is possible to give the telechir more than two pairs of hands and arms and the human controller can control these in turn, clamping one pair before he moves on to the next. This has been proposed for surgery where one pair of hands can act as accurate manually positioned clamps while another pair

---

[†]For micro operation, force feedback should probably be scaled up as the square of the length scale ratio to correspond to areas for forces such as cutting. On the other hand for arms n times larger than the human ones the forces should be scaled down as $1/n^3$ since the weight or mass handled is relevant.

works with forceps and scissors or needle. It has also been proposed (Fig.
8.4) (R. Goertz, 'Manipulator systems development at ANL', *Proc. 12th
Conference on Remote Systems Technology,* Nov. 1964, p. 117) to use one
pair of arms or even two pairs for climbing up scaffolding or holding on to
an underwater structure while another pair of arms and hands carries out
the work on the structure.

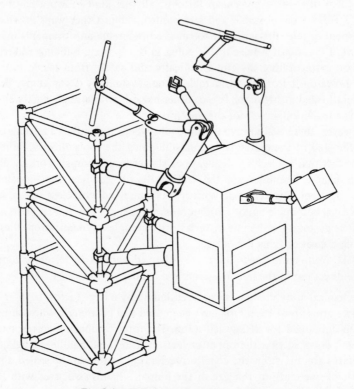

Fig. 8.4 – Telechir concept with six limbs.

(3) The most usual feedback system for telechirs is three-dimensional vision
and force feedback to the human operator's muscles but it is also possible
to give other forms of feedback in terms of signals on a second television
screen or dial gauges so that considerably more information is available for
the operator. There can also be audio feedback either of sounds in the
machine area or of warning signals, triggered for example by tactile contact.

However, feedback is very expensive and the unilateral telechiric system with the
feedforward signals only for the arm and hand is very much cheaper but of
course visual feedback is essential. Moreover, force feedback being much more

complex has much more to go wrong. There is no doubt that mass production of telechirs will bring down the cost of force feedback but it will also probably always be necessary to have only a certain limited amount of force feedback, the minimum probably being the inertial and gravity forces felt at the shoulder and some kind of touch sensitivity at the hand.

During the 30 years since the first work on telechirs, considerable progress has been made in increasing the skill of carrying out operations remotely; thus R. Goertz writing in November 1964 at the Proceedings of the 12th Conference on Remote System Technology after 15 years work on the development of general purpose manipulators for radioactivity says 'Although considerable progress has been made in the development of manipulators the rate of typical hot laboratory work with a high-performance electric or mechanical slave manipulator is only about one-eigth the speed of performing the same operation directly with the hands'. By the proceedings of the 21st conference in 1973, Vertut *et al.* (p. 38, 'Contribution to defining a dexterity factor for manipulators') defined the time efficiency factor as the ratio of the time needed to perform a certain task using a manipulator and the time to do it directly by hand. They studied two types of tasks:

(1) Grasping, load displacement and simple assembly operations.
(2) Turning valves and plugging electric cables.

They found time efficiencies for bilateral master/slave systems ranging from 1.3 to 4.5 while open systems with no force feedback had time efficiencies ranging from more than 100 to 10. For more complex operations the time efficiencies of even the bilateral manipulators range from 3 to 7 and the unilateral range from 30 upwards. In general they concluded that force feedback improves the time efficiency by a factor of 5 to 7 compared with the unilateral arm. In 1977 Wilt *et al*[†]. of General Electric studied tasks which would be too heavy to do by a man directly, in order to compare the *replica master* and *resolved motion rate control* systems with force ratio 24 and size ratio 6.2:1. The replica master has bilateral force feedback. This work showed the replica master to have a distinct improvement in time although this was not as great as with low gain manipulators. The bilateral control makes it easier to perform the tasks without errors, reduces the mental effort and requires a less skilled operator. The force feedback does cause physical fatigue but as he is only handling a small part of the real force this is much more than offset by the reduction in mental effort.

8.2 TELECHIRS TO INCREASE A MAN'S STRENGTH

Two types of telechirs have been developed whereby a man is close to the task and uses direct vision to operate hands with the strength of a crane and with

[†]D. R. Wilt, D. L. Pieper, A. S. Frank and G. G. Glenn, 'An evaluation of control modes in high gain manipulator systems', *Mech. and Machine Theory*, 1977, **12,** p. 373.

force feedback. The first of these is the General Electric Hardiman (Fig. 8.5). This is an exoskeleton into which the man fits and it carries attachment to his body at the feet, forearms and waist. It is capable of load-handling tasks such as walking, lifting, climbing, and pushing with a lift capacity of three-quarters of a ton. The mechanical exoskeleton is controlled by the spatial correspondence to the movements of the limbs of the man inside it and he has a very much scaled-down force feedback so that he feels as if he is manipulating a comfortably light load. The machine is driven with a hydraulic servo system using oil at a pressure of 3,000 psi and the hydraulic bilateral system discussed in Chapter 4. A preliminary series of experiments was carried out at Cornell Aeronautical Laboratory to study the necessary movements of the joints so as to keep the

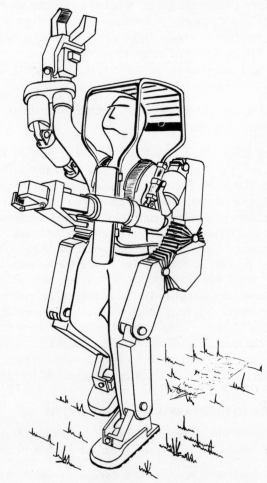

Fig. 8.5 – General Electric Hardiman.

number of joints to a minimum and to study the ranges of movements and the dynamic responses.

The second type of machine amplifier is a kind of combination of forklift truck and crane but with a gripper hand with force feedback to the human controller's hand in place of the hook or lifting prongs. An example is shown in Fig. 8.6 also from a General Electric publication of the 1960s where it is known as the Boom Handler. This uses the same principle of electro-hydraulic manipulation with bilateral force feedback. If developed fully a man could pick up a packing case weighing a ton with the same skill and sensitivity as if he were picking up a matchbox between finger and thumb. GE manufactures industrial manipulators ranging from 100 lb at 4 ft reach to 4000 lb at 24 ft reach (Wilt *et al., Mech. and Machine Theory,* 1977, **12**, p. 374).

Fig. 8.6 – General Electric boom material handler.

8.3 TELECHIR DEVELOPMENT FOR HANDLING OF RADIOACTIVE MATERIALS

When work on radioactivity became essential and dangerous at the end of World War II, people started to develop mechanical master/slave manipulators whereby a human operator was linked mechanically from a master handle to a slave tong so that he could look through a thick protective window into a 'cave' and manipulate highly radioactive materials such as the elements of a nuclear reactor pile. Figure 8.7 shows a sophisticated development of this type.[†] It is necessary to link the two systems through a hole in the protecting wall in the cave well out of alignment with the operator and all kinds of linkages have been provided, the most recent being based primarily on tension wires or tapes, with a pair being necessary for each motion. The problems of mechanical master/slave manipulators are the obvious ones of (1) looseness and play, (2) friction, (3) gravity forces which have to be neutralised with counterweights, (4) inertia of the arm itself and (5) limited volume of coverage. The force feedback is provided by the direct mechanical linkage but this is rendered very insensitive especially by friction.

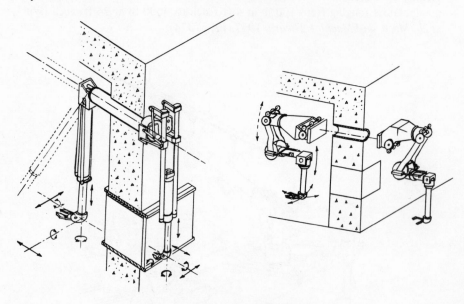

Fig. 8.7 – Mechanically linked telechirs.

In the 1950s and early 1960s the USA was planning to develop a nuclear-propelled rocket and also to develop an aircraft nuclear propulsion engine. The shield of this engine was between the nuclear system in the nose of the plane and the people behind. Hence when on the ground they could not be approached directly for repair and the General Electric Company with other groups, built a vehicle known as the 'Beetle' (*10th Hot Laboratory Proceedings ANS*, November 1962, p. 167). This vehicle weighed 85 tons and moved on tank treads to carry a man in a heavily shielded cab to drive it and operate these manipulators. He was shielded with 0.30 m of lead and five leaded windows 0.6 thick. It carried two power-manipulated telechiric manipulators and the cabin for the man and the

[†]'Manipulator systems development at ANL', R. Goertz, *Proc. 12th Conf. on Remote Systems Technology,* Nov. 1964, p. 117.

two manipulators could be raised on four telescopic hydraulic arms to some 25 ft above ground level. The motions of the manipulator arms are shown in Fig. 8.8. The device had a 550 hp main engine, an auxiliary power package and heavy duty batteries for short distance movement. When fully extended the arms had a reach of 5 m and could support a 40 kg load with a deflection of approximately 25 mm. There was no force feedback but direct viewing through the windows and by means of a periscope.

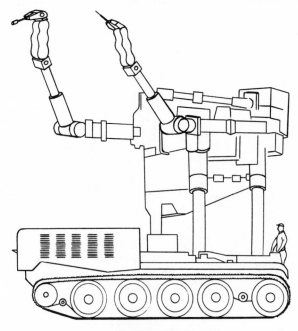

Fig. 8.8 – The 'Beetle'.

Very much more work has been done on the development of master/slave electric and hydraulic manipulators for work in radioactive situations than for other applications. These may be regarded as a development of the mechanical hands used in caves and indeed the first series of prototypes, that associated with Ray Goertz, operated the slave arm and hands by means of tension cables or tapes; but instead of these cables being directly connected to the master arm to transmit its movements they are operated by electric motors driven by servo systems with position comparison from the master arm. The communication was electric signals, which in all the early versions were analogue signals, although work is now starting on digital communication. By 1966 he had reached the Mark E4A ('ANL Mark E4A electric master/slave manipulator', R. Goertz, J. Grimson, C. Potts, D. Mingesz and F. Forster, *Proceedings 14th Conference on Remote Systems Technology*, 1966, p. 115). Each of the seven independent motions of the manipulator was driven by a separate force-reflecting servo and the operator could feel all the load of the slave or half or one-fifth of it. The master arm had a maximum load of $4\frac{1}{2}$ kg and the slave arm of 23 kg in any direction. Figure 8.9 is an outline drawing of the pair (a) slave arms and (b) master arms. Figure 8.10 shows the way in which the seven motors are operated by cables from drums driven by servo masters with about 40 to 1 gear ratio. One motor on each drive unit has an a.c. tachometer and another motor has a geared synchro to provide position signal data. The master and servo drive units

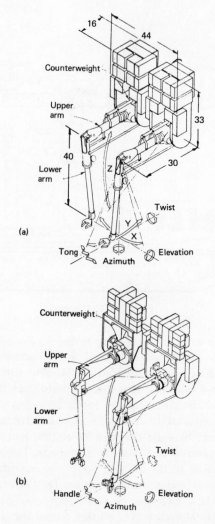

Fig. 8.9 – Outline drawings of (a) a pair of slave arms; (b) a pair of master arms.

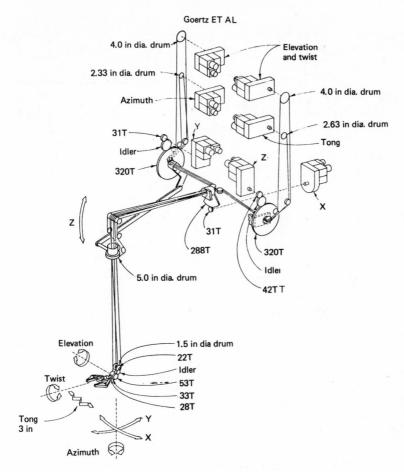

Fig. 8.10 – Servo motor drive system for manipulator of Fig. 8.9.

are connected to each other to provide force reflection but there is a built-in force ratio selector. The servo block diagram of this bilateral system is shown in Fig. 8.11.

Several other lines of work have followed this direction pioneered by Goertz. In 1966 at the 17th Conference on Remote Systems Technology (p. 154, 'Compact servo master/slave manipulator with optimized communication links') C. R. Flatau described a compact servo manipulator in which d.c. servo motors and other servo components were placed within the manipulator arms. He developed a variable force ratio feedback which varied the forces reflected at the master from 1:1 at low force to 3:1 at maximum force. Force reflection is gained by having bilateral symmetry between slave and master as shown in

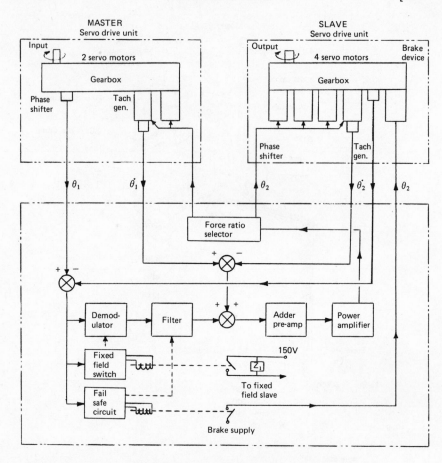

Fig. 8.11 – Block diagram of control system for Fig. 8.9.

Fig. 8.12. By 1977 at the 25th Conference (p. 169, 'A new compact servo master slave manipulator') Flateau has returned to a counterbalanced assembly in which the motors and gear reduction units are placed as counterbalances on the shoulder and the drive is by means of cables. Flatau's work has also been carried on in conjunction with the French Atomic Energy Commission, particularly Jean Vertut, and has led to the Virgule remotely operated vehicle which runs on three wheels and has two arms (Fig. 8.13) (*Proc. 25th Conference on Remote Systems Technology,* 1976, p. 175, 'The MA23 bilateral servomanipulator system', J. Vertut and P. Marchal, G. Debric, M. Petit, D. Francois and P. Coiffet.) These are also based on the use of tape and cable transmission with four of the motor actuators placed in the counterweight mass to give exact balancing, the other three motors are in a fixed assembly which drives the shoulder and elbow motions. The system operates on low inertia d.c. servo motors driven by an amplifier driver system by position control. The system of these arms is now

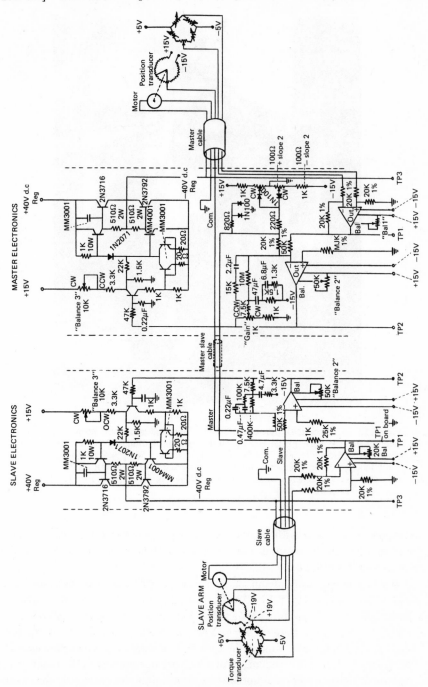

Fig. 8.12 – Bilateral telechir system with force reflection.

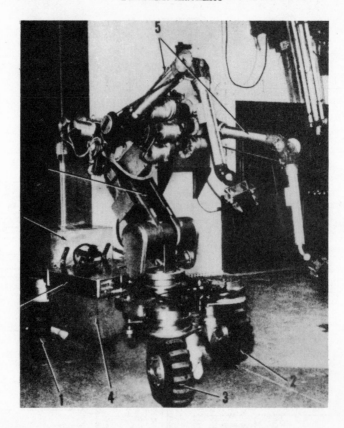

Fig. 8.13 – The Virgule tele-operator. (1) Four self-contained propulsion and steering wheels with special tread-pattern for stair-climbing. (2) Extended front wheel (both extended give stability). (3) Retracted front wheel (both retracted allow passage through a narrow door). (4) Batteries. (5) Pair of MA 22. (6) Three-degree-of-freedom-articulated support. (7) Right-arm power amplifier. (8) Multiplex communication.

commercially available with bilateral symmetry giving the force feedback and the system has also been applied in a deep submergence tele-operator which will be described in section 8.5.

Another line of work stemming from the work of Ray Goertz has been the Italian Mascot system Fig. 8.14 ('An electronically controlled servo manipulator', S. Barabaschi, S. Cammarata, C. Mancini, A. Pulacci and F. Roncaglia, *Proc. 12th Conference on Remote Systems Technology*, p. 143). This has a three-wheeled trolley, electric hydraulic control, a servo TV camera, with three-dimensional rotation and 7 degrees of freedom, all of which are operated by cables from servo motors on the body, except that the hanging elbow joints are operated by rods with parallel motion linkage. The body can be raised hydraulically from the base and

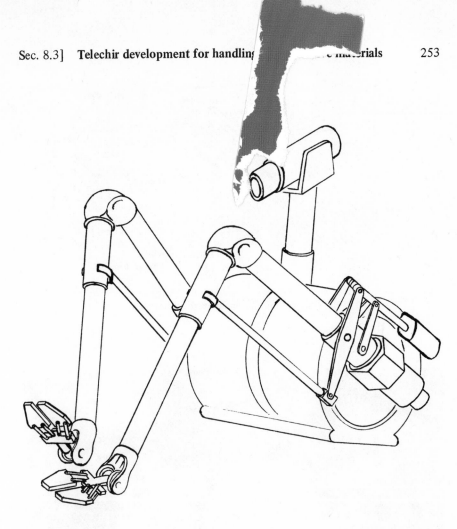

Fig. 8.14 – The Mascot telechir.

the shoulders lifted up so that the hands can operate up to a considerable height. This machine has also been developed for commercial manufacture and in the *23rd Conference on Remote Systems Technology,* 1975 (p. 247), its use on an overhead carriage with two-dimensional movements to cover a large area for a CERN 26-GeV proton syncroton was discussed (R. A. Horne and M. Ellefsplass, 'Long-range, high-speed remote handling at the CERN 26-Gev Proton Synchroton'). There is a bilateral force feedback but only two-dimensional vision from the television cameras; two of them are on articulating arms (Fig. 8.15).

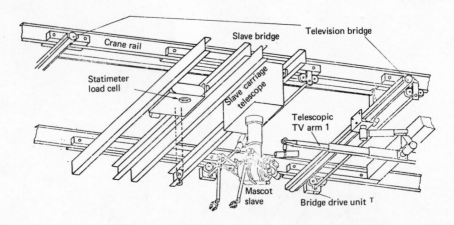

Fig. 8.15 – The Mascot telechir on an overhead carriage.

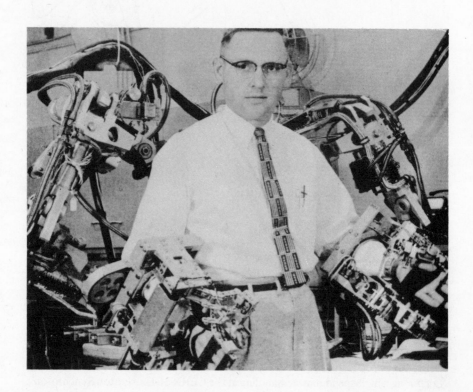

Fig. 8.16(a) – The Handiman servo manipulator: Master station.

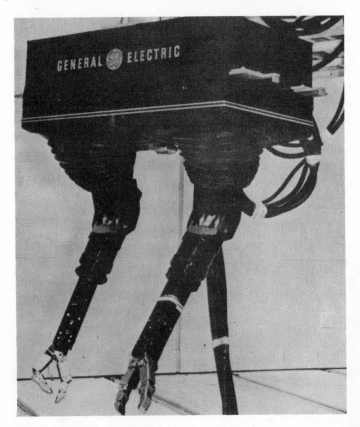

Fig. 8.16(b) — The Handiman servo manipulator: Slave station.

The other main line of work on radioactive handling is the unit based on hydraulic bilateral symmetry. This work largely started from the Handiman developed by R. S. Mosher of General Electric. Figure 8.16 shows the master and slave stations ('An electro hydraulic bilateral servo manipulator', R. S. Mosher, *General Electric Report;* also *SAE,* Jan. 9, 1967). The master and slave only have to be connected electrically since each has its own hydraulic pump, the circuit is that described in section 5.4.2. Each slave arm is capable of lifting 34 kg in its weakest position, that is when it is supported from the other arm by the maximum distance of 2.8 m. There are 10 degrees of freedom for the hand/ arm assembly. The slave to master force ratio can be varied from 10 to 1 down to zero, but it is usually kept at 3 to 1. It is claimed that the use of the hydraulic system enables large force output to be obtained with relatively slight friction, rapid speed of response and comparatively small tubes for the power. The compliance is $3\frac{1}{2}$% of full stroke under maximum load.

A water hydraulic telechiric manipulator was described by K. B. Wilson (Fig. 8.17) ('Servo arm – a water hydraulic master/slave manipulator', *23rd Conference on Remote Systems Technology*, 1975, p. 233). This also has bilateral symmetry of the servo mechanisms and uses *terminal control* in which the movement of the master's hand operates an arm of quite different geometry to his arm but of the same geometry of the slave. It is possible to get friction so low that the force threshold is 115 g. They developed a special back-drivable servovalve with poppet and piston construction rather than conventional spools to make it more dirt tolerant and they developed a compact water pump to supply the 7.6 l/min water for each arm at 100 bar. The system uses a microcomputer and A/D conversion hardware for counterbalancing and can therefore be readily made to work on a computer program.

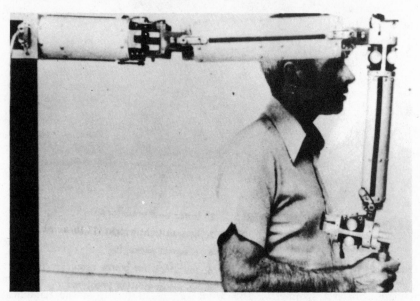

Fig. 8.17 – A water hydraulic telechir manipulator.

Unilateral manipulators have been developed in Germany and Britain and the Mobot in California. The very considerable need connected with the nuclear energy programme for handling these highly dangerous materials has clearly led to an immense amount of work and expenditure.

8.4 TELECHIRS FOR WORK IN SPACE

Much of the earlier work on telechirs for use in space was based on a control operator on the ground and this system suffers from the fundamental difficulty that the sensory message from the telechir to the human and the return signals telling the telechir what to do can take many seconds or minutes at the great distances involved in space work, although it is only a fraction of a second for orbits close to the earth. A remotely controlled tug-based device intended for exchanging modules on orbiting spacecraft is described by F. C. Runge ('Space tug/spacecraft/module exchanger', *Mech. and Machine Theory,* 1977, **12,** p. 451). This is to change modules held in a 5 × 5 block, when one of them breaks down or when it is required to change a scientific or working pay-load. It is particularly for communication and earth observation satellites, both for geosynchronous orbits and lower orbits. The manipulators are fairly simple as it is only necessary to select the appropriate module in the block, unlatch it and replace it with the correct new one once the block is latched on to the spacecraft. The servicing of communication satellites in geosynchronous orbit for design failures, random failures, and wear-out failures is also discussed by G. D. Gordon ('A user assessment of servicing in geostationary orbit', *Mech. and Machine Theory,* 1977, **12,** p. 463).

A third type of orbital servicing module system where the modules are arranged in concentric circles and handled by a pivoting arm is discussed by G. W. Smith and W. L. D. Rocher ('Orbital servicing and remotely manned systems, *Mech. and Machine Theory,* 1977, **12,** p. 65). M. H. Kaplan and A. A. Nadkarni ('Control and stability problems of remote orbital capture', *Mech. and Machine Theory,* 1977, **12,** p. 57) discuss a proposal for a free-flying 'teleoperator' launched from a shuttle vehicle. The problem is to retrieve passive spinning and precessing satellites by means of giant fingers on an arm. The torque component on the fingers is absorbed by the grip of the fingers to produce a smaller spin motion of the combined system. The arm and hand are spun to match the expected satellite movement as it grips and there are countermasses which articulate to maintain dynamic balance.

In 1978 two papers in the *Proceedings of the 26th Conference on Remote Systems Technology* were concerned with manned remote work stations (MRWS) ('Manipulators for large-scaled construction in outer space', C. A. Nathan and C. R. Flatau; and 'Large scale manipulator for space shuttle payload handling', M. J. Taylor, p. 90).

The system described by Nathan and Flatau (Fig. 8.18) consists of a universal cabin to carry one man and the cabin is planned to have two sophisticated manipulator arms placed on a stabiliser arm with which it can grip hold of the work. The cabin is too small for the man to work full-sized controls and his arm controls are thus scaled down to one-third of the external arms. The drive and control is done with the system of electric motors with cables to the joints, and there are 7 degrees of freedom on each arm. This crew cabin with its three arms has windows for direct vision of the task being done although it can also have television cameras. It is 7.1 m in diameter with a 1 m degrees hatch, top and bottom and is 2.5 m high. This can be carried on a big arm from the shuttle or it could be free-flying or run on rails attached to the task. The ultimate aim

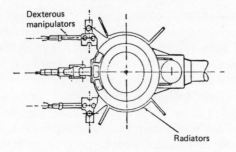

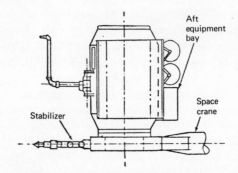

Fig. 8.18 – One-man space cabin with telechirs.

is that it should be possible to construct large structures in space from shuttles after 1990. Taylor (Fig. 8.19) discusses the shuttle remote manipulator system in which the manipulator arm is some 15 m long and is attached to the shuttle so that it can carry an end-effector capable of capturing a payload with quite large misalignment and also position payloads relative to orbiter axis with precision. The arm is controlled by a human operator on the shuttle who has direct vision from two windows looking aft into the shuttle cargo bay and also from two windows above him. He also has two television screens located adjacent to his control hand which receive signals from cameras located on the payload bulkheads and the wrist of the manipulator arm. There is no gravity problem in shuttle orbit and the arm can handle a 14.5 ton payload and give it a maximum speed of 0.06 m/sec. The operator controls the arm by means of two 3-degrees-of-freedom joysticks, the left-hand joystick provides the three transitional motions of the manipulator while the right-hand one gives the three rotational movements. The control system consists of one servo for each degree of freedom of the joints. Since the end-effector may have a very high or a very low inertia according to whether it is carrying the load or not, the gearing for motors on the drive is a combined low speed and high speed gear group. The low speed planetary gear gives 85% efficiency so that is adequately back-drivable to give force feedback.

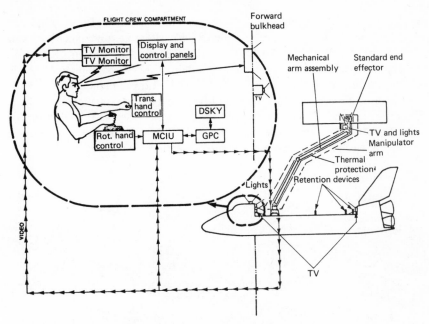

Fig. 8.19 – Shuttle telechir system.

8.5 UNDERSEA TELECHIRS

Three types of undersea telechiric manipulators have been extensively developed, particularly for naval purposes but also for commercial purposes such as surveying and maintenance of oil pipelines under the sea or of structural rigs for drilling. A third purpose for which undersea unmanned vehicles have been used is the scientific one of surveying the sea bottom in great detail, but these do not have telechiric slave arms.

(1) The first type is the small, manned, free-swimming submersible which may carry telechiric arms and have windows for direct vision. It may also have a lock system for loading a diver in and out. It must be powered with batteries or oxygen/fuel combustion and hence its work capacity and duration undersea are strictly limited.

(2) Towed submersibles with human operators inside. The operator usually looks through a window at the manipulator arms. American examples are the Alvin, Seacliff, Turtle and Scarab, and there is also the Dutch Bruker manipulator system attached to the Mermaid submersible. This is a hydraulically operated pair of arms controlled by an operator looking through a hemispherical glass window and each arm has 6 degrees of freedom with a hand with parallel grip.

(3) The unmanned systems operated by a cable from a ship. These include the
     American CURV (Fig. 8.20) and RUWS (Remote unmanned working
     system), the British Angus, the French ERIC and the Russian CRAB ('An
     overview of non-US underwater remotely manned manipulators', A. B.
     Rechnitzer, *Mech. and Machine Theory,* 1977, **12**, p. 51) built for operation
     down to 4000 m with a 7 degrees of freedom telechiric manipulator. This is
     free-swimming but can also rest on the bottom of the sea. The latest version
     is the MANTA which causes the operator's chair to follow the pitch and role
     of the telechir to increase the operator's handling ability.

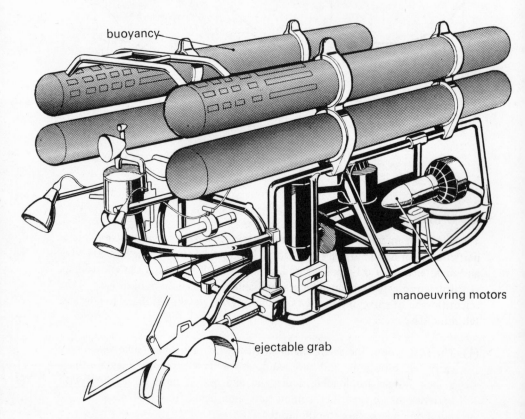

Fig. 8.20 – US CURV (1967) undersea telechir.

Under the first category of manned, free-swimming submersibles an example
is the Vickers Pisces. This is battery-operated with lead acid batteries and has
two men in it and two arms, one sophisticated manipulator arm and one torpedo
recovery arm. The life support period is from 24 to 100 hours ('Mobility on the
sea floor', *Engineering,* April 1974, p. 271). The big problem with such a machine

is to know exactly where it is. This is done by means of a scanning sonar and a transducer below the keel of the support ship. Pisces I recovered a tug sunk in 650 ft of water when it had been equipped with a hydraulic chain cutter on its manipulator which could cut through a 15 mm steel chain. More recently they have built a glass-reinforced, plastic submersible (*Offshore Engineer,* June 1977, p. 37). The Angus ('A Navigable General Purpose Under-water Surveyor', R. M. Dunbar and R. T. Holmes, *Electronics and Power,* 17 April 1975) is a small unmanned submersible tethered to a ship to give detailed information about the bottom of the sea. Electric power comes in a cable from the ship at 415 volts and propulsion is by means of a three-phase four-pole squirrel cage induction motor, voltage variation being used for speed control. The main cable is 18 mm multicore polyurethane sheathed with a breaking strength of one ton force. This carries three power conductors and 40 control and instrumentation conductors with one coaxial cable. Navigation is done by the long baseline acoustic technique with two transponder buoys on the sea bed. Information is sent back in the form of a closed circuit television and wide band hydrophone and there is also a 16 mm cine camera.

The US navy remote unmanned work system RUWS ('Position and force feedback give manipulator precise control, ease of operation', R. W. Uhrich and A. E. Munson, *Hydraulics and Pneumatics,* September 1973, p. 178) has a heavy duty claw with 4 degrees of freedom and a manipulator with 7 degrees of freedom weighing 28 kg and capable of lifting 20 kg at the maximum reach of 1.3 m. The whole system is hydraulically operated. The hydraulic system is so arranged that it works on ambient pressure plus 69 bar; the ambient pressure operates through that part of the 3 mm bore nylon pipes which are exposed to ambient pressure. They have developed a four-way pressure control servovalve and seven of these for the manipulator are mounted on a double-sided aluminium manifold head in an oil bath. The oil filled external tubing also carries the potentiometer cables but these are internal on the final arm. The six actuators of the arm and wrist are rotary vane and the parallel movement of the grip is driven by a cylinder, so the system is $\widehat{RR}\ \overline{RRRR}$ P. The three wrist rotations operate on axes going through a point and have only unilateral control. The three arm movements are bilateral and the grasp is an open-ended control system giving a grip force proportional to the trigger depression against a spring.

The development of a system of manipulator arms and hydraulic tools which can be attached to this and other US Navy deep ocean manned systems for salvage operations is described in a paper by Estabrook *et al.* ('Development of deep ocean work system', N. Estabrook, H. Wheeler, D. Upler and D. Hackman, *Mech. and Machine Theory,* 1977, **12**, p. 569). The Work Systems Package is a complete unit which has two grabber arms which can secure and hold on to a work piece for stability or assist the dextrous work arm. These are hydraulically actuated, have 6 degees of feeedom and a lift capacity of 114 kg at 2.74 m extension, with a grip force of 410 kg. The dextrous work manipulator is a 7 degrees of freedom hydraulic actuated rate controlled arm. The tubular aluminium holder which can carry a dozen tools is positioned opposite to the primary manipulator and just out of view of the frontal viewing area and bits such as drills and sockets for these tools can be held in clips at the edge of the tool holder. There is a high flow hydraulic system for powering tools as well as a low flow system for operating manipulators and the television cameras (pan and tilt) and a winch. Among the tools available are a chipping hammer with a rotary motor driving a cam against a compression spring, a low-speed rotary tool and piston-actuated cable cutter, spreader and jack.

SCARAB — The Submersible Craft Assisting Repair and Burial developed by MBA ('Design and application of remote manipulator systems', C. Witham, A. Fabert and A. L. Foote, *Proceedings 26th Conference on Remote Systems Technology,* 1978, p. 76) has been developed for undersea telephone cable repair and surveillance (Fig. 8.21). It can locate, unbury, attach, cut and return a cable down to 1830 m depth. It has an arm with 4 degrees of freedom ($\overline{\text{RRRP}}$) plus a tool which it carries. They have concluded that variable rate control is

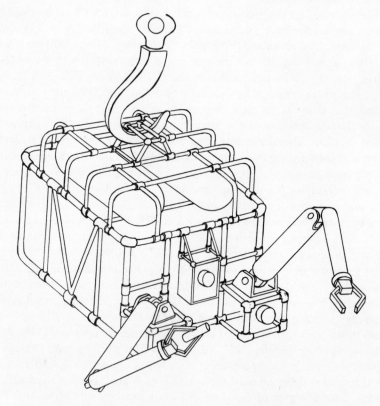

Fig. 8.21 — US SCARAB underwater telechir.

cheaper than position servo-control but is, of course, subject to drift so they use the latter to avoid drift. The actuators are hydraulically powered rotary vane and cylinder.

An unmanned manipulator submersible developed for the French Navy has been described in some detail by J. Charles and J. Vertut ('Cable controlled deep submergence teleoperator system', *Mech. and Machine Theory,* 1977, **12,** p. 481). This is designed to search and investigage on the sea floor down to a depth of 6000 m. It consists of a 'fish-house' called PAGODE which acts as a lift between the bottom and the surface and carries the main cable and the 300 m tether cable for the neutral bouyancy teleoperator 'fish' which is called ERIC II (see Fig. 8.22). The combined system is dropped to the

Fig. 8.22 – French ERIC underwater telechir.

required depth and then ERIC swims out of the 'fish house' to carry out the task. It is planned to have ultimately 6 degrees of freedom for the television camera, controlled by the rotation and movement of the operator's head in the support ship, the movements being in relation to his chair fixed in space. The telechir 'head' has binaural microphones connected to two earphones on the helmet in which the operator's head is placed (see Fig. 8.23). This helmet has binocular TV display and is counterweighted. ERIC weighs 4–5 tons and has 100 kW propulsive power supplied at 600 volts 400 Hz to keep the voltage control systems fairly constant and to transmit the required signals. The data transmission system is a composite one with analogue for specific purposes and digital for general purposes. The position of the 'fish'

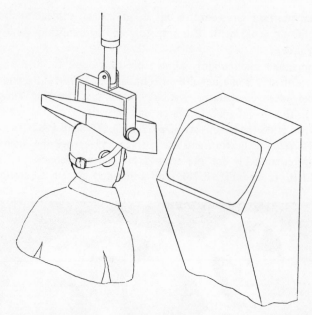

Fig. 8.23 — Head mobility system for ERIC telechir.

is determined over large areas by bottom acoustic transponders and panoramic sonar and locally by sonar and forward TV cameras. The basic intention is to mimic the overall capacity of a human diver without the limitations imposed by pressure. The system has a main propeller gimballed so that it always thrusts along the tiller tension direction with a tether so that it swims like a free body. It has three pairs of ducted propellers with variable blade angles and with thrust transducers for three-dimensional steering and trim. The two bilateral arms are worked by an electromechanical system with cable transmission for the movements.

8.6 TELECHIRIC MINING

The main cost of extracting solid minerals of all kinds from the ground is the cost of making holes in the ground in which human beings can work safely. With liquids or gases one has only to drill a hole of less than $\frac{1}{2}$ m diameter and release or pump the fuel through the hole so these are much cheaper to win. With coal or other solid minerals on the other hand one has to make the mine safe for human beings to work in, have passages large enough for them to travel along, pump fresh air down and keep the content of methane in the atmosphere below the combustible limit; one has to pump water out and install lighting systems and man transport. Conditions of underground working are becoming less and less acceptable as human working standards rise and it is no longer possible to win coal from seams as thin as 0.5 m where a hundred years ago men

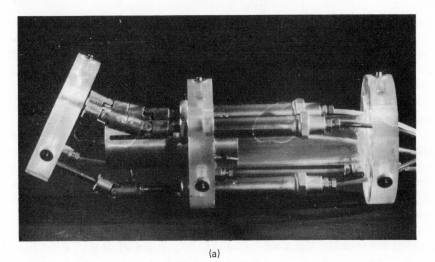

(a)

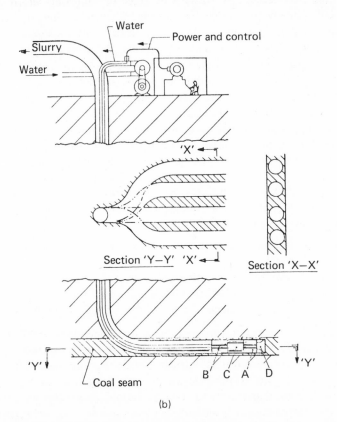

(b)

Fig. 8.24 – Steerable Mole Miner. (a) Model. (b) Diagram of operation.

used to lie on their side and break the coal out with a pick. If we can find a way of bringing coal to the surface without men going underground it can give the world a reasonable energy supply for at least one hundred years, including all the people in the under-developed countries. The author has been working on this problem for 20 years, his first proposal ('Mole mining', in Crookes and Thring, *Energy and Humanity,* Peter Peregrinus, p. 166) was a surface-controlled mole miner in which the coal crushed by the circular cutting head was pumped to the surface in a stream of water. This was similar to an oil well drill except that the mole could be steered round a corner and steered in the seam of coal by having sensing devices in the cutter at the four corners, so that it could tell when it reached harder rock outside and the human on the surface could then steer it. Figure 8.24 shows a model and a diagram of operation. The cutting force was produced by dividing the body at the front of the flexible support tube in to two parts, first the rear part was jammed into the hole by three feet (each $120°$ of the cylinder) that were forced radially outward while the cutting head on the front part was forced forward by three or four arms which could be worked differentially to provide the steering. When it reached the end of the stroke the radial feet on the rear part were released and the radial feet on the front part were forced out so that by reversing the stroke of the three main rams the rear part and the flexible tubes could be drawn up behind it. The main problems with this machine would be (1) the friction of the tubes dragged behind it and (2) supporting the bore hole so that the machine could be withdrawn when it reached the end of the cut.

Another scheme which has been proposed is to have a mole which seals itself in the hole it burns through the coal seam so that when it is fed with air at one atmosphere and it compresses this to 5 or 10 atmospheres it can throw a high pressure jet of air at the coal in front. The burnt gases pass through a turbine on the machine which drives the compressor and and electric generator and the electricity and cooled combustion gases are brought to the surface. The problems in developing this machine are:

(1) sealing the machine in the seam as it slowly advances with a pressure difference of many atmospheres;
(2) dealing with ash from combustion and shale in the seam;
(3) it requires fully developed telechirs as men could not go down the pit for repairs, maintenance or changing the region being burnt once it had been lit.

    Considerable experiments have gone on underground gasification ('The underground gasification of coal' *National Coal Board. Pitman,* 1964; reappraisal 1976). But it has been shown by calculation that it is extremely difficult to do more than drive off the volatiles leaving the coke behind; the gas has a very low calorific value (less than one tenth that of natural gas) unless pure oxygen is used for combustion and it is difficult to pierce another hole for further gasification after one hole is exhausted as this requires sending men down into a burning pit. In any case coal has far too many uses on the surface to confine it to a gas which is only fit for burning for power at the pit head, indeed underground gasification was only considered in Britain for coal near the surface but uneconomic for mining conventionally. However, it may be possible to develop a method of underground distillation or steam extraction for the tar in the enormous tar sand deposits of Athabaska (Alberta, N. Canada), preferably using telechirs.

In the process of pipe jacking concrete cylinders are thrust hydraulically into a horizontal hole in the ground while one man at the front end of the hole excavates the ground material so that the hooded shield at the front can be pushed forward while another man barrows the spoil back to the entry pit. This is clearly an ideal application for telechirics since the adaptability of the man excavating to varying soil conditions (e.g. finding large boulders in clay) is an essential prerequisite of the operation.

In 1970 the author proposed true Telechiric Mining (see Fig. 8.25) ('Mining without men going underground', M. W. Thring, 1980 *ASME*, 81-Pet-21) which

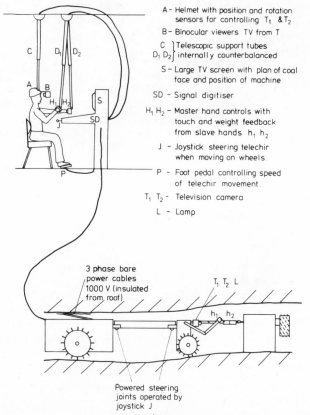

A - Helmet with position and rotation sensors for controlling T_1 & T_2

B - Binocular viewers TV from T

C — Telescopic support tubes
D_1 D_2 internally counterbalanced

S - Large TV screen with plan of coal face and position of machine

SD - Signal digitiser

H_1 H_2 - Master hand controls with touch and weight feedback from slave hands h_1, h_2

J - Joystick steering telechir when moving on wheels

P - Foot pedal controlling speed of telechir movement

T_1 T_2 - Television camera

L - Lamp

3 phase bare power cables 1000 V (insulated from roof)

T_1 T_2 L

h_1 h_2

Powered steering joints operated by joystick J

Fig. 8.25 — Diagram of telechiric mining.

may be defined as a system in which the underground machinery for cutting roadways, cutting the coal, transporting it, and for maintaining the conditions in the pit, are all essentially similar to those which are at present operated by men down a modern well-automated pit, but the men stay on the surface *never going underground* and do the tasks that they would at present do underground by means of mobile telechirs communicating with their hands, arms and eyes, combined with extra sensing devices on machines such as coal cutters. These telechirs may be independent, mobile machines, equivalent to the head and shoulders and arms of a man attached to a body which can move at a brisk speed in a space in which the man can only crawl on hands and knees. Figure 8.26 shows a series of models of possible telechirs for mining. Alternatively they may be attached as arms and eyes on a machine so that a man on the

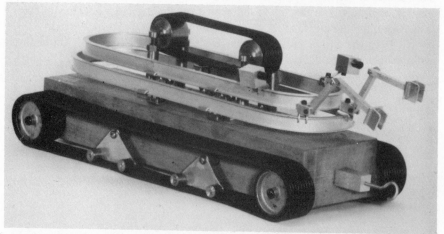

Fig. 8.26 – Models of mine telechirs.

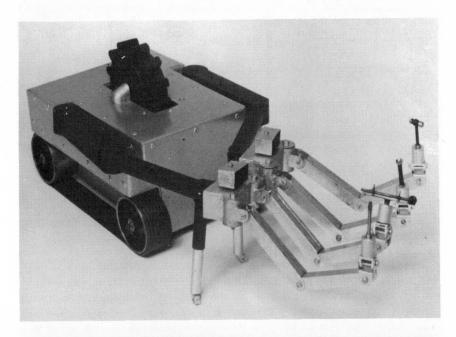

Fig. 8.26 — Models of mine telechirs.

Fig. 8.26 – Models of mine telechirs.

surface could for example change a pick on a coal cutter by means of the arm attached to the coal cutter itself, which was normally not in use. A single mobile telechir can be operated at different times from the surface by various specialists such as a maintenance and repair mechanic, a mining engineer, a geologist, or a machine operator. For routine operation the machine operator on the surface would receive all the necessary information from an automated coal cutter and override the automation when this information showed the necessity or if the automatic steering was out of order. When he observed a fault he would call the repair mechanic to take his place at the telechir controls.

Ultimately, therefore, we can envisage coal mines which contain no human life support system because no human being even goes down; they would therefore not be ventilated so that it would no longer be necessary to have a complex dual system of passages and galleries for the throughflow of air connected to two separate shafts and it would be possible to run the mine under sealed conditions so that the atmosphere was entirely mine gas with no oxygen, no fear of explosion of fire and the mine gas could actually be burned (see Fig. 8.27). A whole mine could be operated from a single drift and since it is not necessary to convey men to the face every day the underground area worked could extend several tens of miles from the drift and it would be possible to win coal far under the sea from a drift at the land. It is also possible to avoid surface tips either by having a washery at the surface and automatic return of the washery refuse as a back fill in the working area or even by having automatic coal separation underground and not bringing the dirt to the surface at all (see Fig. 8.28). If this can be developed it becomes possible to mine seams heavily banded with other material. Surface subsidence has already been eliminated in mines in

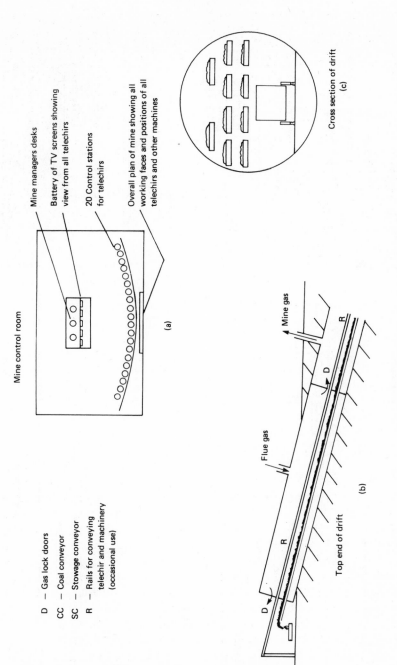

Mine control room

Mine managers desks

Battery of TV screens showing
view from all telechirs

20 Control stations
for telechirs

Overall plan of mine showing all
working faces and positions of all
telechirs and other machines

(a)

Cross section of drift
(c)

Mine gas

Flue gas

Top end of drift

(b)

D — Gas lock doors
CC — Coal conveyor
SC — Stowage conveyor
R — Rails for conveying
      telechir and machinery
      (occasional use)

Fig. 8.27 – Possible telechiric mine layout.

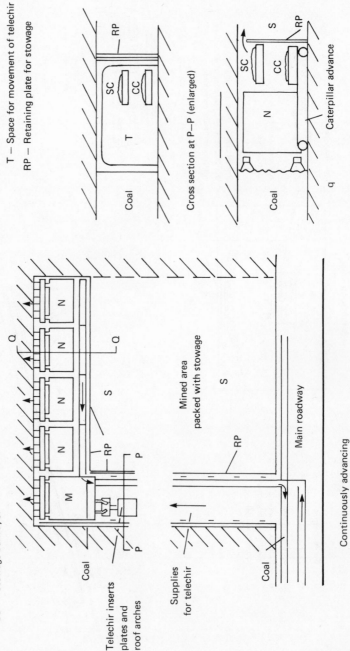

M – Roadway coal cutter (chain picks) with attached telechir for own repairs

N – Linked coal cutters with attached telechir for own repairs

CC – Coal conveyor
SC – Stowage conveyor

S – Stowage
T – Space for movement of telechir
RP – Retaining plate for stowage

Cross section at P–P (enlarged)

Caterpillar advance

Cross section at Q–Q (enlarged)

Telechir inserts
plates and
roof arches

Supplies
for telechir

Mined area
packed with stowage

Main roadway

Continuously advancing
longwall coal face

Fig. 8.28 – Possible telechiric mine layout details.

Poland by pumping a sand—water slurry back into the worked area so that the bottom one-third of a thick coal seam can be worked and then afterwards the second and third (top) sections of the seam are worked with the operation supported on the sand filling the lower cuts. Other possibilities are the working of seams ultimately down to 0.3 m thick, the working of seams at depth far too hot for humans and the working of very steeply sloping seams. It may also be possible in the future to develop telechiric mining systems that could work in a completely flooded pit.

We can envisage the control room of a mine (Fig. 8.27(a)) of the future with 20 telechir control stations, each occupied normally by a machine operator but with the television images from each one being available for other experts and the manager, while a large diagram on the wall shows the position of each telechir in the pit and the situation and status of all other machines.

'Some opportunities in tele-operated mining' (E. R. Palowitch, *Mech. and Machine Theory,* 1977, p.493) is a discussion of the possibility of setting up an automated long wall mining system by means of a double ranging drum long wall shearer. When operated by hand this machine has two operators, one ranges the leading drum at the roof level and the other ranges the following drum at the floor level. The trajectory is determined by the line of the chocks and they must also control the roll in order to follow the seam. Coal beds may thicken or thin in short distances, undulate and are sometimes faulted, eroded or wanting. The most important problem is to sense the interface between the coal and the rock above and below. The operators rely on the noise and vibration of the cutters and visual observation of the cuttings, the previous cut and bands of different coloured coal. It is also necessary for them to check that the roof falls regularly behind the chocks as they advance since otherwise excessive stresses may be caused on the chocks. The ideal system would involve pitch control automatically from a sensor while a remote human would select goals, plan, monitor production, detect failures and determine maintenance needs. The ideal measurement for pitch control would be the location of the coal/rock interface prior to cutting but this is so far unachievable. It is however, possible to locate the interface during cutting, either by the use of sensitised picks which determine the force on the cutting elements or by listening to the difference in the machine's vibration. Both these methods, however, are more effective when the coal seam lies under much harder rock. If it is bounded by softer shale they are less effective. The NCB at Bretby have developed a nucleonic probe based on the measurement of the internal gamma ray activity of the shale behind the coal and with this it is possible to adjust the ranging drum automatically. This operates 3 ft behind the cutting drum and can be made to leave several inches of coal under the roof, which is required if the rock above is too soft to arch. The authors concluded that it would be necessary to have a well balanced mixture of men in remote control and automatic control systems and that the man must be provided with video observation, microphones for auditory information, some type of accelerometer to give him information on the position and movement of the machine and a number of dial gauges for check out and monitoring.

Figure 8.29 shows a mine materials manipulator for working underground in thick coal seams built in the USA ('Design and application of remote manipulator systems', C. Witham, A. Fakert and A. L. Foote, *Proceedings 26th Conference on Remote Systems Technology,* 1978, p. 76). The human operator lies in it under a protecting roof with direct visual observation of the task and controls the manipulator arm by moving the handle on the control arm (Fig. 8.29(b)). Five of the six arm degrees of freedom are position—position servos (without force feedback) but the elevation is rate controlled by a lever on the control arm. Grip is controlled by a foot pedal. The arm has 2.08 m reach and 118 kg lift capacity. To reduce the danger of damage to the hydraulic lines operating the joint movements the electro hydraulic servovalves are close to the joint actuators so that only two hydraulic lines, pressure and return, are needed for the whole arm. The system is battery-operated.

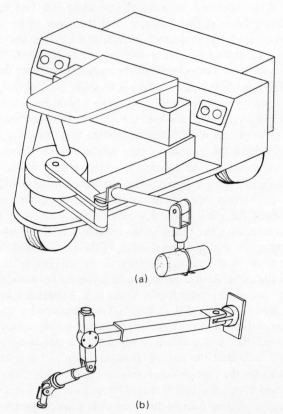

(a)

(b)

Fig. 8.29 – MBA mine materials telechir. (a) Manipulator. (B) Replica master control.

8.7 TELECHIRIC SURGERY

Thring ('Perspective', *The Blue Cross Magazine,* First Quarter, 1972, p. 27)
(Fig. 8.30) has considered the possibility of telechiric surgery in which a surgeon

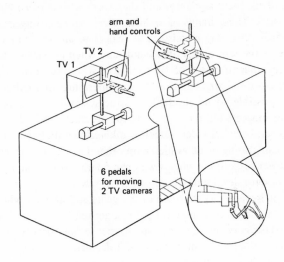

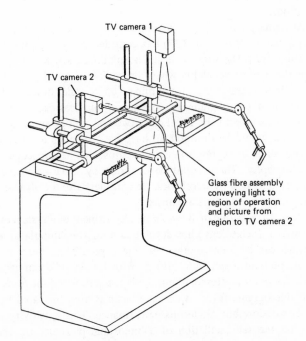

Fig. 8.30 – Surgeon's telechir.

sits at a control panel outside the sterile enclosure in which is the patient and several pairs of telechiric hands and any tools and equipment needed. The surgeon can use these pairs of hands one after another by connecting control mittens to a different terminal for each; thus he can move one pair of hands into a certain place and fix them as clamps and then work with another pair carrying different instruments. These hands of stainless steel can be stronger and thinner than his fingers or he can plug in scissors or scalpel to an arm, one set of hands can be one-tenth of the size of his own hands and when he uses these the video system would give a magnification of 10, so that he would see and feel as though his hands were the normal size but that he was operating on a patient 10 times as big. It is also possible to use multifibre optic systems for seeing round the back or inside an inaccessible part, such as the stomach with the same video screen, or he can have several video screens looking from different angles. What the surgeon sees can also be made available on other TV screens for instructional purposes or for other surgeons or nurses who can operate other hands.

One of the modern developments of surgery is sewing back severed hands or fingers in which it is necessary to sew (1) the veins and arteries which may be less then half a millimeter in diameter, (2) the main tendons to work the thumb and fingers and (3) the nerve bundles together so that sensory nerves are connected to sensory and motor nerves to motor nerves. Telechiric surgery for this type of operation could clearly be of much value in relieving the strain on the surgeons in such operations.

A. D. Alexander ('Impacts of telemation on modern society', *1st CISM-IFToMM Symposium,* Vol. 2, p. 122) has considered the possibility of surgery at great distances from the surgeon in his hospital (see Fig. 8.31). This could be life saving in the case of accidents where immediate surgical help can make all the difference or in cases at sea, down mines, or in under-developed countries. It can also be used to up-grade the quality of medical care and uses scarce medical people more effectively. In all cases two-way colour and voice TV would be essential but it would be desirable to have basic diagnostic equipment at the emergency site such as a respirometer, stethoscope, fluoroscope and image intensifier radiography. These could be operated by a para-medic on the spot or even remotely operated by telechirics. He considers also the ultimate possibiliy of remote treatment by master/slave arms.

Umetami (p. 12, *Resume of Work 1980,* Department of Physical Engineering, Tokyo Institute of Technology) has developed a micro-manipulator with tactile sensibility which can be used to grasp an object, pick it up, move it, and stab a hypodermic needle and inject fluid into it. With forceps this can handle objects of size from a few micrometres up to hundreds of microns and the force sensitivity is 10 to 100 micrograms force. A ceramic beam is used for rapid firm motion and a polymer for slow but dexterous performance. This sensitivity is obtained by the fact that the self-oscillation of a bimorphic piezoelectric beam is proportional to the contact force.

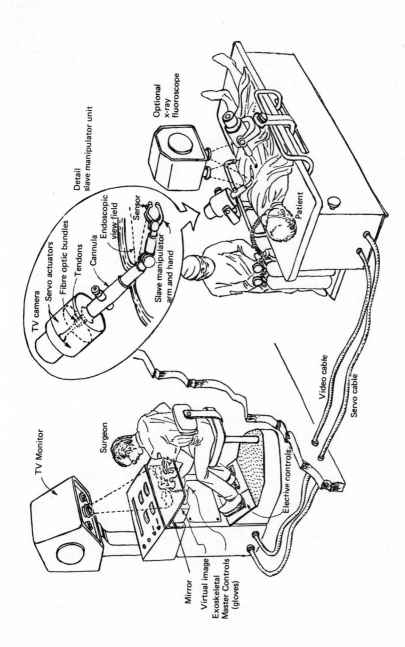

Fig. 8.31 — Remote surgery concept.

Development of a telechiric micro-manipulator which can be used without special training to give positioning resolution of μm and a frequency bandwidth of 10 Hz described at the 4th CISM-IFToMM Symposium 1981 (p. 251, 'Servo micro-manipulator "Tiny-Micro Mk. I" ', K. Marsushima and H. Koyanagi). The operator holds a control shaped like a pencil connected by a control arm (see Fig. 8.32) to a fixed point as he would hold a scalpel and the slave arm

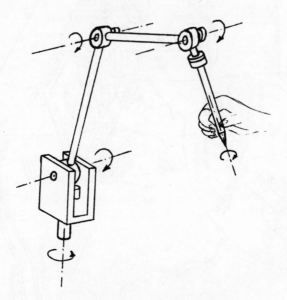

Fig. 8.32 – Control arm of a micro-telechir.

copies the movements of the control arm on a scale reduction of 150:1. All movements are R, the control arm having potentiometer position sensors full scale resolution ± 500 μm while the slave actuators are electrohydraulic with d.c. motors working a screw which pushes a diaphragm which causes a small quantity of silicon oil to flow in a tube and move a diaphragm. The stroke of the slave arm actuators is 10 mm. To reduce the offset caused by the piston torque of the actuator a non-linear compensator and tachometric feedback are used.

CHAPTER 9

The use of Robots, Telechirs and Mechano-chiropods in the Creative Society

9.1 THE FUTURE AND WHERE DO WE GO FROM HERE?

In Chapter 1 I mentioned *the robot fallacy* that it is a good idea to replace workers by robots if the sole advantage is to make the product with less expenditure. This is a fallacy because the cost of human dissatisfaction in throwing a man out of work and the national cost of extra unemployment more than offsets the saving to the firm, so that the use of robots for this purpose is really a Government subsidy to the production cost. Here, however, we have to consider a much bigger fallacy, *the fallacy of perpetual growth*. Just as a living organism grows to a certain size and then levels off with a slight shrinking in old age and ultimate death so industrial organisations, national organisations and even civilisations go through the same stages. During the period of growth and expansion optimistic people believe that there will be continued growth of industrial sophistication and output almost without limit.

It is possible to cite many examples of this fallacy of perpetual growth. Twenty years ago in Britain both the electricity generation industry and the steel industry had plans for a steady growth of an exponential type. In the case of the electricity industry this was expected to continue at a rate as high as 9% p.a. Now they both can produce much more they can sell. Another example is the development of supersonic passenger aircraft based on the assumption that because aircraft speeds had risen steadily they would go on increasing, and pass the sound barrier and some designers even considered that hypersonic flight might become commercially viable. A few years ago people were planning ships of 500,000 tons and more and the only limit to their ideas was the structural strength of the ship. Similarly they were supposing that the amount of goods carried by ships, including oil would increase indefinitely. Now many cargo ships are laid up.

The British coal industry is a classic case. It did expand exponentially up until the start of World War I in 1914 (Fig. 9.1) but since then it has declined in a rather irregular manner from the peak of 270 M tons a year to the present figure of just over 100 M tons a year. The most obvious current example of the fallacy of perpetual growth is in the field of computers and micro-electronics. These very clever and sophisticated devices clearly have many more applications in store; nevertheless, even for them the demand will eventually saturate and many of the enthusiastic firms starting off in this direction will go bankrupt.

There are three main reasons why the steady increase in the *sophistication* of technology will level off.

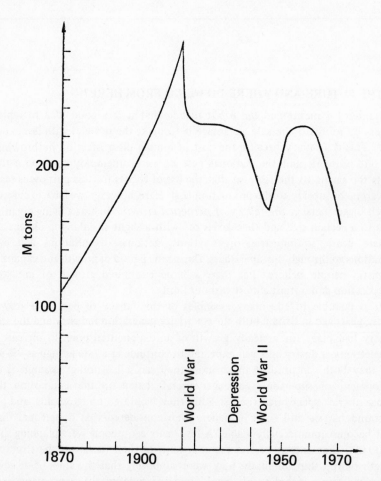

Fig. 9.1 – British coal production over a 100 year period (from H. E. Collins, 'The Revitalized Coal Industries', *Colliery Guardian* publication, 1975).

(1) The urgent needs of the under-developed countries.
(2) Rising unemployment in the developed countries.
(3) Shortage of fuel and other raw materials.

The first of these is vitally important, even to the rich countries, because it is quite certain that the world cannot remain peaceful and avoid World War III with the eventual use of nuclear explosives and neutron emitters, chemical and biological poisons, if the population of the poor countries continues to rise and they continue to get poorer. It is not that they will attack the rich countries directly — they will not be able to afford to buy from the rich countries enough weapons to do that. It is rather that world tension, due to the frustration of desperately poor people, who observe a few rich people in their own country or learn of the wealth of others in rich countries, creates an atmosphere in which revolutions, guerrilla warfare, kidnapping, hijackings and other forms of rioting and violence regularly occur. The great powers get involved in the supply of weapons to one side or the other and the hawks in the great powers find opportunities to try out their latest weapons as the conflict escalates. This is already happening in many parts of the world.

Equally dangerous to the whole world is the situation of an intermediate country where a substantial fraction of the population lives at Western standards and pays taxes high enough to enable the Government to buy the latest weapons and make nuclear explosives from the products of their own power stations. When the other part of the population of the intermediate country, which is living in poverty as desperate as the poorest countries, becomes restive, the Government is tempted to attack its neighbours and become popular on a wave of nationalistic feeling. Sooner or later a country which has its own nuclear bombs will decide to use them — that is the purpose for which it has acquired them.

The conclusion is inevitable that if the rich countries continue to use the poor countries as a source of raw materials which are paid for by exporting sophisticated machinery and consumer goods, then the war which must result will bring all high technology to an end.

Thus *the first requirement for a stable world in the twenty-first century* is that instead of pushing further and further into high technology to give more luxuries for ourselves in the rich countries and spending ever-increasing sums on sophisticated weapons, we must devote enough of our available resources of manpower, equipment and research to developing the machinery that will enable the people of the poor countries to produce a decent standard of living for themselves. It is in our own and our children's and grandchildren's interest that we make this change in the rich countries, since it is stupidly short-sighted to enjoy luxuries now that will cause the destruction of our whole civilisation early in the next century.

This application of our engineering skills to the production of machines that will enable the whole of humanity to grow enough foods and earn homes, clothing, medicines, education without absorbing all their waking energies, has been called by various names. Schumacher introduced the expression 'Intermediate technology' which is accurate in that the degree of sophistication is intermediate between 'western high technology' and 'UDC low technology', but it has been criticised as being somewhat patronising and having the implication that 'high technology' is superior to 'low' or 'intermediate technology'. In any case highly sophisticated technology may often be required for the equipment of the underdeveloped countries for example in providing local power from wind, sun or combustion of agricultural refuse. Other names which have been used are 'radical technology', 'alternative technology' and 'appropriate technology'; of these 'appropriate technology' seems to me preferable because it implies that the new technology will be appropriate in the developed countries as well as in the underdeveloped countries. My own preference is for *humane technology* as it implies that technology is being used for the purposes of humanity and not being pursued for the sake of its own complexity and sophistication.

9.2 RAW MATERIALS AND UNEMPLOYMENT

The argument against the Luddites has always been that the introduction of new technology enabled the existing manufactures to be made more cheaply by fewer workers but at the same time those workers still employed could be paid more real money and they would be in a position to buy more goods and have a rising standard of living. This, in turn, meant that new manufactures could be started and the workers displaced from the old industries could now be employed in the new ones. During the first 200 years of the Industrial Revolution this steadily rising standard of living has indeed provided manufacturing jobs to a rising population and to the former agricultural workers who had moved to the towns and factories from the farms. While the agricultural production per unit land area in the rich countries has increased more than fourfold the proportion of people employed on the land has gone down from more than 50% to 2−4%. However, this high production is achieved by the use of fossil fuels for fertiliser (especially N-fixation), for pesticides and herbicides (necessary because of monocropping of high yield but disease-prone varieties), and for making and operating tractors, combine harvesters and other machinery.

Every engineer, and indeed biologists and all other scientists, know that exponential growth cannot continue for ever. A 4% per annum growth rate means a fifty-fold increase in a century and fifty-fold again in the next century. In point of fact, the energy consumption per capita per annum in Britain has gone up in the last 200 years from about $\frac{1}{2}$ ton of coal equivalent (TCE/c.a.) to 5−6 TCE/c.a., i.e. by a factor of 10−12. The coal equivalent of the food eaten by an active man in a year is about 1/6 TCE/c.a., so we can say that whereas 200 years ago he was served by 2−3 energy slaves (fuel for heating and cooking and animal, wind and water power) now he is served by 30−40 energy slaves.

There are three main factors which will halt this exponential growth.

(1) When one relates the growth of wealth per capita measured as GNP/c.a. (gross national product per head per year) to the energy consumption the relation is astonishingly linear (see Fig. 9.2). However, the GNP can be increased without corresponding increase in energy consumption by (1) capital investment on methods of fuel economy (triggered by rising fossil fuel prices) and (b) the tendency to develop sophisticated electronic gadgets which do not require large amounts of fuel to make them because they do not use large masses of steel unlike cars, ships or refrigerators. Robots on the whole use considerably more energy than humans for the same job as they tend to use force rather than skill and to be slower, just as machine tools use more energy than hand tools per job; but machine tools are immensely faster, more accurate and reliable in repetition.

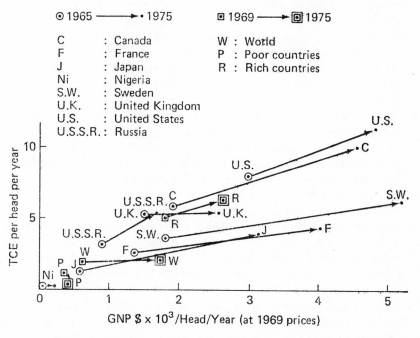

Fig. 9.2 – The relation between energy consumption and GNP.

There is a definite limit, however, to fuel economy; i.e. there comes a point below which any further increase in GNP/c.a. is essentially dependent on an increase in TCE/c.a. Since by far the greater part of the energy on which the industrial revolution is based comes from the fossil fuels (in the US in 1970 oil first, natural gas second, coal third, the renewable fuel hydro power fourth, wood fifth and nuclear power sixth) and the reserves of these fossil fuels are definitely limited, it follows inevitably that as the most easily accessible fossil fuels are exhausted the cost of energy will rise much more rapidly than other costs, e.g. labour costs. In turn this means that the first factor which will cause the exponential growth rate in GNP/c.a., i.e. in individual consumption, to level off in the rich countries will be the steeply rising cost of fuel.

The high fuel cost will also make it essential that consumer goods, such as cars, which require a large amount of fuel to make them will be made to last a human lifetime and this will add to the unemployment among those at present making them. The car industries of the world are already facing a falling market even with the present short-lived versions.

Allied to this fuel factor we have the same effect for many of the other materials which are mined, such as nickel, phosphorus (for fertilisers), manganese, tungsten, zinc, tin, copper and others such as aluminium, iron and silicon which are quite plentiful but require a lot of fuel to reduce them from the oxides. Again, we are taking the cheaply won and more concentrated ores and leaving only the more inaccessible or leaner ones for our grand-children, and this will not only prevent the standard of living in the rich countries from rising but will eventually cause it to fall, as indeed is happening at the present time.

(2) The second main factor which will prevent the employment of people thrown out of work by automation and robots from being employed in manufacturing new products for a rising consumption is space. There is certainly a saturation point for the number of cars on the roads which already been reached in all the major cities of the rich countries. Similarly there is a saturation point for the number of gadgets a household can have in the kitchen or bathroom. Many households have cupboards full of gadgets that are never used.

(3) However, there is a third factor which, although intangible, is as important as the other two. This is the commonsense of the consumer which eventually makes him or her totally resistant to the latest sales gimmick. The *gullibility factor* varies enormously from individual to individual, but some people have refused to have TV and many more will not feel impelled to buy colour video recorders, home computers and video information telex systems. There is indeed a growing movement away from the dependence on gadgetry, as indeed there is towards healthy walking or cycling in place of the use of cars for even the shortest distance.

For these three reasons it is likely that we have already reached the limit to growth of personal spending of real money in the rich countries, as is evidenced by the steady increase in unemployment (see Fig. 9.3) and the incidence of inflation. Inflation must clearly occur whenever people are paid more money but their real purchasing power does not rise. Unemployment has fluctuations of course, due to such factors as Government decisions to spend more on the arms race (it goes down whenever there is a threat of war and the stock market rises) but basically, as shown in Fig. 9.3 (taken from a paper by Sir Alex Smith), it has risen steadily in Britain even since the end of the World War II regardless of which political party was in power.

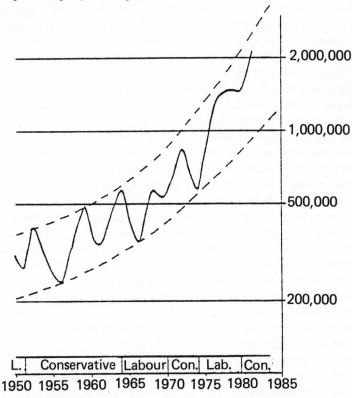

Fig. 9.3 – Unemployment in Britain since World War II (from Sir Alex Smith, The Stanley Lecture, 1980).

The principal argument in favour of a massive increase in the use of robots in countries such as Britain, is that by far the greatest use of robots is in Japan (Heginbotham, *Proc. I. Mech. E.,* 1981, **195**, 37, p. 409, 'Present trends, applications and future prospects for the use of industrial robots') where there were some 6,000 in 1981, while Japan has the lowest unemployment and the most successful export of sophisticated devices of all the rich countries. Six thousand robots might be replacing 40,000 workers if they work three shifts and do twice as much as work as a man in each shift, although this figure for the number of workers replaced is probably considerably on the high side. The population of Japan is some 110 million so that it has at least 30 million adult workers and thus robots at present are displacing about one worker in 1,000 — a very small fraction compared to the number displaced by automation of other kinds.

This argument makes it clear that the Japanese success is not due to their use of robots but is due to 'their ability to provide the right goods at the right cost, at the right time' (Heginbotham, loc. cit., p. 410). Heginbotham concludes: 'the widespread application of industrial robots in a country which is commercially less successfully would not necessarily lead to a similar situation. This does not mean, of course, that we [i.e. Britain] can be complacent and say that we have no need to apply robots but must strive harder to create sales so that we too can obtain the advantages of amplification'.

The short-term economic argument for considerable national expenditure on the development and application of industrial robots is that it will make a country like Britain able to produce more cheaply, export more successfully to pay for its essential imports and so increase employment in increased production.

However, it is clear that no one country can solve its problems alone for if every country tries to 'export their unemployment', all, even the most successful, will eventually suffer disaster, owing to market saturation.

If one considers an industry like the car industry which has based its whole planning on large sales of short-lived cars, with annual changes of models, it is clear that this argument cannot work for any one country because the competition between all the car-manufacturing countries to export to each other and only import raw materials must lead in the end to a situation in which fewer and fewer people can *afford* to buy cars. The poor countries have a few rich people who can afford to buy them but the this does not help significantly either to absorb the over-production of the rich countries or to prevent their own people from starving. As soon as a poor country achieves a certain level of national wealth it wants to make its own cars, while the exploitation of the poor countries to grow foods to export to the rich countries instead of using their agricultural land to feed themselves increases the danger of revolution which can easily escalate to World War III.

So how many robots and telechirs will we really need in the future for the whole world?

9.3 THE LONG-TERM PROSPECT

When one looks five or even ten years ahead one obtains very different answers to all industrial and economic questions to those obtained when one looks 20—30 years ahead to the beginning of the twenty-first century. The young engineers we are training now, hope to have a 40-year career and therefore it is essential to give them the longer perspective.

The particular economic and industrial question we are trying to answer here is 'what is the long-term future of industrial robots and telechirs?', but in order to answer even this specific question it is essential to consider the much bigger question 'what are the necessary conditions for a stable world in the twenty-first century?'.

I have tried to answer this question fully in my earlier books, *Man, Machines and Tomorrow,* (Routledge and Kegan Paul, 1974), *Machines, Masters or Slaves of Man?* (Peter Peregrinus, 1974), *The Engineers' Conscience* (Northgate, 1980), but here it is necessary to summarise the argument briefly for it leads to an answer which illustrates Thring's Economic Principle (*Engineers' Conscience,* p. 128), 'Whatever is right for my grandchildren is always uneconomic now and almost always impolitic!'

If we are to have a stable world in the twenty-first century it is essential to get rid of the terrifying risk of nuclear war and the ever-increasing expenditure[†] of the Earth's limited resources on the production of weapons for which our greatest hope is that they will never be used for the purpose for which they are produced. Even to dispose of them when they are superseded is incredibily difficult and dangerous. To achieve this world peace, certain conditions *must* be fulfilled.

(1) At present the population of the poor countries is about 2,700 million, about twice that of the countries that have had the benefit of the industrial revolution. However, the population of the poor countries is increasing at 3% p.a. and that of the rich countries at less than 1%, so that in one generation there will be four times as many people in the poor countries as in the rich. This, of course, leads to the growth of deserts and a rapid worsening of the proportion of seriously undernourished people in the world.

It follows from this difference in population growth rate between rich and poor countries that there is one humane way of levelling off the world's population and this is *to provide the benefits of the rich countries to all people: namely a fully adequate standard of living and education.* This must be done within one generation if we are to avoid total disaster which will spread contagiously to the rich countries.

(2) The second necessary condition is *essentially to eliminate the enormous differences in standard of living and use of resources between the ordinary people of the developed and underdeveloped nations.* Only in this way can the world tensions due to ordinary people of the poor countries seeing the wealth of a few in their own country and hearing of the wealth of the many in the rich countries, be eliminated.

(3) The third conclusion follows from the second taken in conjunction with the problems of exhaustion of resources, rising unemployment and mismanagement of the land for short-term gain discussed above. These factors mean

---

[†]Governments of the world spend three times as much on military activities as they do on health, twice as much as on education, and 30–40 times as much as the rich countries provide as aid to the poor countries (*Engineers' Conscience,* p. 33); so if the rich countries were to divert 10% of their military expenditure to give real aid to the poor countries they would increase the latter 4–5-fold.

that it is quite impossible for the doubled world population of 8,000 million people who will exist in 40 years (unless of course we kill most of them with World War III before then) to live at a level which uses the world resources as fast as we use them at present in the rich countries. To take energy as an example, the average use in the rich countries is about equal to the British figure of $5\frac{1}{2}$ TCE/c.a.; the average for the whole world is about 1.8 TCE/c.a. and that for the poor countries about 0.5 TCE/c.a. If twice the world's present population used three times as much energy per head as the present average then the world consumption would have to rise by a factor of six from the present which is clearly an impossible figure because of the exhaustion of easily won fossil fuels and the greenhouse effect due to CO_2 from combustion.

Therefore the third necessary condition for a peaceful world is: *the consumption of raw materials and energy per capita in the developed countries must fall to about one-third of the present figure (one-sixth in N. America) and that in underdeveloped countries must rise to 3–5 times the present figure in the next 40 years.*

This is the conclusion which illustrates Thring's Economic Principle; for any democratic politician in a rich country who asked for a 66% reduction in consumption of energy and raw materials over the next 40 years would certainly not win election, while a dictatorship of left or right takes even more short-sighted decisions based on dogma, militarism or fear of revolution. However, if one calls it a 3–4% reduction in raw materials consumption every year for 40 years and if enough intelligent people see that this is essential for the long-term survival of our civilisation then there is some hope.

In the next section of this chapter I shall assume that we do arrive at a stable world in the next century by taking the above steps and try to evaluate the role of robots, telechirs and medical mechano-chiropods in such a society.

9.4 THE ROLE OF ROBOTS IN THE CREATIVE SOCIETY

The motive force behind the industrial revolution was individual profit and such a motive, tempered by social and legal controls in a mixed economy, was reasonably adequate as long as the economy and the standard of living of the individual was continually expanding. However, now that the limits of this continual expansion in the rich countries have been reached and because of the danger of impingement of the desperate poverty of the poor countries on the peace of the whole world, the short-term motive of maximising immediate profit is no longer sufficient for the survival of our civilisation.

A stable civilisation in a world where the material cake is insufficient to give everyone all the goods that millions of people have become accustomed to, must necessarily be based on a motive or measure of personal success totally different from status symbols. The *motive of creative self-fulfilment* is much more gratifying in the long run to the individual and has the advantage that the more one person achieves this objective the more is available for others. For this reason we could call the stable world society which is the only alternative to disaster in the twenty-first century, the Creative Society.

The essential character of the Creative Society is that everyone in the whole world can earn a perfectly adequate standard of living for themselves and educa-tion for their children by doing a job which is not excessively boring or dangerous and does not absorb more than half their adult life-energies. As far as possible they will find creative satisfaction in their work (as in teaching, health care, craftsman-ship and artistic and musical activities) but they will also have enough leisure energy to widen their creative self-fulfilment to a point where it gives real meaning to their lives.

It is not feasible to suppose that the problem of unemployment will be solved in the Creative Society by everyone working very much shorter hours, for two fundamental reasons connected with human motivation. Holidays could get somewhat longer, working life could be shortened to cover the ages say 20 years old to 55 years old and the working hours of the week could be reduced a little; but it is certainly not feasible to reduce the present working life which is approximately 45 years × 35 hours a week[†] × 48 weeks in the year = 75,600 hours to less than one half this figure.

The first reason is that the conscience of a normal human being tells him or her that they owe a fair measure of work to repay the advantages and products of civilisation. One of the principal causes of the breakdown of the Roman Empire was the fact that the citizens of Rome came to expect bread and circuses without working for them. People need to feel that they are useful to society and society needs them.

The second reason is that very few people have enough self-discipline to live happily without a framework of required activity to their lives. Most un-employed people find the time hangs very heavily on their hands and sometimes they even have difficulty in finding somewhere to go. It takes a great deal of courage and determination to find for oneself useful and interesting activities to fill an unemployed life, however large one's unemployment pay, and this can cause as much psychological and even physiological harm to the individual as the feeling that one is of no use.

The objectives of industrial production become quite different in such a society, just as they are different in a poor country which wants to employ as many people as possible on a project. If we have cars they will have to be built to last 40–50 years to save fuel in manufacture, and probably will be all alike and certainly will use less than half as much fuel as the most economical ones do at present. Conservation, recycling and use of renewable energy resources will

[†]It was 47 when I was a boy.

have top priority. Many goods that are at present made entirely in mass-production factories will be largely handmade, where the result is more satisfying and it is more economical in raw materials and fuel to do so. Examples are furniture, clothes, children's toys, and tools, instruments for artists and musicians. On the other hand, sophisticated products like refrigerators, cars, light bulbs and TV sets will still be made by automated factories. When cars are made to last a lifetime far fewer people will be employed in the robot-using operations of making them and far more people will be engaged in the interesting task of diagnosing and repairing faults. Similarly when fuel economy is really seen to be essential we shall repair and maintain old buildings, sewage systems and railways and, because every job is different, there will not be jobs for robots.

In a Creative Society recycling of all metals and other household and factory refuse and of P and K, and even eventually soluble nitrogen in city night soil, will be essential.

The function of robots will not be to replace humans primarily in order to make goods more cheaply but will be

(1) to do all those jobs which are not worthy of humans because they are so boring and repetitive that they do not make use of human skill of hand-eye coordination, do not require the adaptability of a human to an unexpected deviation, and can be learnt in minutes with a repetition cycle of a few minutes;
(2) to do jobs which are uncomfortable or dangerous, for example because of heat, vapour, or droplets, or require complex manipulative skills for a heavy tool but are repetitive.

It is doubtful whether it will be worth the enormous expense of developing industrial robots beyond the second generation (where they have limited adaptability to expected variations in the surrounding conditions) because jobs which require proper skills of trained hand—eye coordination, such as allowing for knots in wood when doing joinery, wood carving, loosening a rusted bolt without shearing it off, weeding a flower bed, or cooking a really tasty meal will be sufficiently interesting for human satisfaction.

All these factors lead to the conclusion that first and second generation robots will play a significant role in the Creative Society in making and assembling components of certain mass-production articles and in doing routine maintenance jobs, but that the number per industrial worker will probably not rise to more than twice the present Japanese figure, so that the total world market with a world population of 8000 million might be as high as half a million but is likely to be considerably less.

It is very unlikely that robots will play a significant role in agriculture because the energy limitations will prevent the low manpower, high mechanisation systems of the West from spreading to the poorer countries as they become richer; and the need for full employment in a stable world will probably lead to a system of compromise in all countries. In such a system some 20—40% of the working population will work on the land; at present in the rich countries only 2—4% do so while in the poor ones the proportion is of the order of 70—90%. In any case even routine jobs on the land such as hoeing weeds or transplanting require considerable human skill and there is more satisfaction in making things grow than in the repetitive industrial jobs which are better done by robots.

9.5 THE ROLE OF TELECHIRS IN THE CREATIVE SOCIETY

When we turn to the use of telechirs in the Creative Society the picture is much clearer, because the principal aims are to enable the human craftsman to use all his trained skills either (1) with his body remote from a dangerous, inaccessible or uncomfortable work zone or (2) to multiply his strength without loss of manipulative ability or (3) to enable him to do microscopic work as though the object and the tools were of a size enlarged to a more convenient one.

While the use of telechirs for mining minerals will eventually enable a miner to win 3—4 times as much coal or other mineral with a day's work as at present, and so fewer miners will be employed for the same output, there is no doubt that all the underground minerals that are needed in the Creative Society will be won by these means because (1) working underground in a mine is necessarily uncomfortable, unhealthy and dangerous compared with the alternative jobs which will be available on the surface, and (2) the additional minerals that can be won will be essential to provide the good things of the industrial revolution to all 8,000 million peoples. Eventually in such a society even the minerals and coal available to telechiric mining will be exhausted. When this happens, if we can remain peaceful that long, then all metals and other non-fuel elements will have to be totally recycled, while only renewable fuels such as agricultural or forestry by-products (e.g. straw and branches of trees), solar and wind energy, hydro power, wave and tidal energy will be available.

In the meantime, however, it will be necessary to have perhaps 250,000 mining telechirs in the world. They will probably first be developed for the very deep mines such as the S. African gold mines or for coal seams which are uneconomic to operate with men in them, such as very thin, deep or steeply sloping mines.

Underwater telechirs similarly will be very important indeed in the creative society, because diving in connection with undersea oil wells is becoming increasingly hazardous as drilling occurs at greater depths in oceans and as manned submersibles with telechiric hands are limited by the battery capacity. Eventually we can envisage a well being drilled and capped and pipes being connected to it entirely by machinery situated on the sea bottom and operated by telechirs controlled by a man in a ship or on the shore. Once the use of platforms with legs, or floating platforms which have to be located exactly, is eliminated the whole problem becomes much simpler because the cables to the telechirs which carry the power and communication signals will not foul legs or floats and the ship carrying the telechir operators need only be anchored approximately. As the underwater vision by sonar or high intensity light plus TV cameras (very close up) has led to workable solutions, this equipment can be produced in large numbers as readily as we now produce cars, TV sets or computers.

Another field in which telechirs will be essential in the Creative Society is in dangerous or uncomfortable work in factories, such as handling explosives or detonators, rebricking or patching a hot furnace or maintaining chemical plant carrying poisonous or explosive gases.

No doubt there will still be accidental fires in houses, factories, warehouses, cars and aeroplanes and a walking telechir will be controlled by a fireman standing back to use a fire extinguisher right at the heart of the fire (where much less damage will be done than by flooding water in from a distant hose nozzle), or even possibly to rescue human beings.

The use of strength-amplifying telechirs to enable a man to handle a one-ton package as carefully and accurately as if he were holding a matchbox will also follow as soon as the problems of tactile and force feedback to the man's arm and hand have been solved in a way that can be mass produced.

The micro telechir will be extensively used in surgery (e.g. eye surgery) and in micro-electronics.

9.6 SCEPTROLOGY IN CREATIVE SOCIETY

Since we have postulated that there would be no wars or muggings in a creative society in which people judge their success in life by their creative self-fulfilment and everyone can have a worthwhile job which earns them and their family enough of everything needed for a satisfactory life, there should not be so many limbless people. Similarly in such a society since the emphasis will be almost entirely on preventive medicine rather than curing the illness after it has occurred so there should be relatively few amputations for surgical reasons and relatively few children born with such defects as spina bifida or brittle bones and relatively little incidence of illnesses such as multiple sclerosis. Nevertheless, for the few cases of such need, the work on powered artificial arms, hands and legs which is being pursued by caring engineers all over the world, will still be useful and will be funded as liberally as weaponry is at present.

Author Index

Subject Index